Chemistry
for the Applied Sciences

Chemistry
for the Applied Sciences

BY

W. STEEDMAN, PH.D., A.R.I.C., A.M.INST.F.
Lecturer in Organic Chemistry, Heriot-Watt University, Edinburgh

I. H. ANDERSON, PH.D., A.R.I.C.
Lecturer in Inorganic Chemistry, Heriot-Watt University, Edinburgh

R. B. SNADDEN, B.SC., A.R.I.C.
Lecturer in Physical Chemistry, Heriot-Watt University, Edinburgh

PERGAMON PRESS
Oxford · New York · Toronto
Sydney · Braunschweig

Pergamon Press Ltd., Headington Hill Hall, Oxford
Pergamon Press Inc., Maxwell House, Fairview Park,
Elmsford, New York 10523
Pergamon of Canada Ltd., 207 Queen's Quay West, Toronto 1
Pergamon Press (Aust.) Pty. Ltd., 19a Boundary Street,
Rushcutters Bay, N.S.W. 2011, Australia
Vieweg & Sohn GmbH, Burgplatz 1, Braunschweig

First edition 1970
Library of Congress Catalog Card No. 77-112287

*Filmset by the European Printing Corporation Limited, Dublin, Ireland
Printed in Great Britain by A. Wheaton and Co. Exeter.*

Contents

Contents

Preface

AT A time when the disadvantages of over-specialization in science and technology are becoming more clearly recognizable there is little need to defend the practice of teaching chemistry to prospective graduates in engineering and other applied sciences. The desirability of interpreting the behaviour of a large variety of materials in terms of their physical and chemical structures is acknowledged by the emergence of materials science as a discipline in itself. It is our experience that such an interpretation is possible only if a student has some reasonable acquaintance with chemical principles. This book attempts to set out these principles and to examine as far as it can within its scope their significance in the applied sciences.

The book divides naturally into two parts, the first part being mainly general physical and inorganic chemistry, the second organic chemistry. The early Chapters, 1–6, contain accounts of fundamental topics concerned with chemical bonding and structure. In Chapters 7–18 these concepts are developed and their relevance to a number of topics in applied science examined. Chapters 19–23 constitute a short, non-mechanistic introduction to organic chemistry, permitting in Chapters 24–28 a study of a number of important organic materials. Throughout, the accent is on the application of chemical principles rather than process descriptions. The book is designed to meet the varied needs of undergraduate engineers whose chemical backgrounds vary widely. Thus, while we hope that much of the material will stimulate the interest of the undergraduate with a good G.C.E. "A" level pass, we hope too that there is sufficient fundamental material to allow those with lower entry qualifications in chemistry to acquire a sound grounding in the subject.

We have not used SI units since it is still widespread practice in chemistry both in Europe and more particularly in North America to employ traditional systems of units.

The material embodied in this book forms the basis of a first course in chemistry given to undergraduates in engineering sciences, chemical engineering, applied physics and applied mathematics in this University.

Heriot-Watt University, Edinburgh   R. B. S.
                                                       I. H. A.
                                                       W. S.

CHAPTER 1

Atomic Structure and Bonding

1.1. Atomic theory

The idea that all matter is composed of minute discrete particles, or atoms, has existed for hundreds of years, but it was not until the nineteenth century that this idea became a quantitative theory in accord with known experimental facts. It was at this time that Dalton advanced his theory which postulated that

  (i) matter is composed of small particles called atoms,

  (ii) atoms can neither be created nor destroyed,

  (iii) atoms of one particular element are all exactly alike, but different from the atoms of all other elements,

  (iv) chemical combination takes place between whole numbers of atoms in definite numerical proportions.

Although we now know that not all of these ideas are strictly correct, they are the foundations on which modern atomic theory has been built. While a detailed study of the structure of the atom lies more in the realm of physics, some knowledge is essential for a reasonable understanding of how atoms bond together.

1.2. Subatomic particles

It is now accepted that the atom is not the ultimate particle, but is composed of even smaller entities. A fairly large number of subatomic particles have been discovered to date, but only three are considered to be stable and only these are of direct interest to the chemist. These particles are the *proton, neutron* and *electron.*

Protons and neutrons have about the same mass but differ in that a proton carries unit positive charge while the neutron is electrically neutral. The mass of an electron is about 1/1850 that of a proton and the electron carries unit negative charge. It can be seen therefore that practically all of the mass of an atom is accounted for by protons and neutrons, and since atoms are electrically neutral the number of protons and electrons in any given atom must be equal. The next problem we must consider is how these individual particles are arranged within the actual atom.

1.3. The nuclear atom

Whereas it was once believed that atoms had a homogeneous structure, it is now considered that the atom consists of a small positively charged *nucleus* containing the neutrons and protons, and that this is surrounded by a relatively diffuse arrangement of electrons. The diameter of the nucleus is of the order of 1.0×10^{-12} cm, while that of the atom as a whole is about 1.0×10^{-8} cm. The nucleus is held together by attractive forces operating between protons and neutrons, and there is a tendency for these particles to exist in equal numbers within the nucleus. The structure and stability of an atomic nucleus is considered in more detail in Chapter 2.

Every atom is characterized by its *atomic number* (Z), which is the number of protons contained in its nucleus. Hydrogen, the simplest element, which has a nucleus consisting of only one proton, has an atomic number $Z = 1$. Helium, the next simplest element, whose nucleus is composed of two protons and two neutrons, has an atomic number $Z = 2$.

So far we have a picture fairly similar to that described by Dalton, with atoms characterized by their atomic number and all atoms of any particular element having the same mass. We find, however, that this is not always the case. Hydrogen, for example, can exist in three different forms; hydrogen with a nucleus containing one proton, hydrogen with a nucleus composed of one proton and one neutron (*deuterium* or 'heavy' *hydrogen*), and hydrogen with a nucleus containing one proton and two neutrons (*tritium*). These atoms of hydrogen, containing the same number of protons but different numbers of neutrons, are called *isotopes* of hydrogen and obviously the mass of each isotope differs from that of the others.

Many of the elements can exist in two or more isotopic forms, and this has in the past created some difficulties in the definition of atomic mass. Since the mass of an atom is extremely small it is convenient to define a relative atomic mass scale, choosing one element as a reference. Prior to 1961 the reference element was oxygen which was given a mass value of 16 atomic mass units (a.m.u.). Since naturally occurring oxygen contains three isotopes, each containing eight protons together with eight, nine and ten neutrons respectively, a certain amount of ambiguity existed because chemists ascribed a mass of 16 a.m.u. to the average mixture of isotopes, whilst physicists referred the scale to the single isotope containing eight protons and eight neutrons. Now confusion is avoided by assigning 12 a.m.u. to the isotope of carbon containing six protons and six neutrons.

On this scale the relative masses of protons, neutrons and electrons are 1·00728, 1·00867 and 0·000549 a.m.u. respectively. It would be wrong, however, to assume that the atomic mass of any element can be calculated exactly simply by adding together the relative masses of its constituent parts. In the formation of a nucleus from individual protons and neutrons, a small proportion of the total mass is converted into energy. The consequences of this phenomenon are discussed more fully in Chapter 2.

1.4. Electronic structure

In general terms the structure of the nucleus does not have any marked effect on chemical behaviour. Thus we find only very slight differences in chemical reactivity between the various isotopes of oxygen. It is true that differences are rather more pronounced when we consider the chemical behaviour of hydrogen and deuterium, because in this case the addition of only one neutron to the nucleus doubles the mass of the atom. This is the exception rather than the rule, and it is the electronic structure of atoms we must investigate if we wish to understand their chemical behaviour.

Modern theories of atomic structure are highly mathematical rather than pictorial, and although they are more useful than the older theories they are correspondingly more difficult to understand. One of the basic postulates of modern theory (Heisenberg's Uncertainty Principle) is that it is impossible to define both the position and the energy of an electron simultaneously, since the process of attempting to measure the location or momentum of an electron will itself have some effect on its position or energy. The behaviour of an electron is therefore described in terms of the *probability* of finding an electron at any particular point in space. We can regard an electron in two different ways, either as a discrete particle which will spend most of its time in regions of high probability or alternatively as a cloud of negative charge with a high density in high probability regions and a low density in low probability areas. For example, we could picture a hydrogen atom as in Fig. 1.1, where the single electron of the hydrogen atom is represented as a diffuse cloud of negative charge.

This diagram describes two aspects of the behaviour of the electron; first that the electron is most likely to be found in the regions close to the nucleus where the density of shading is

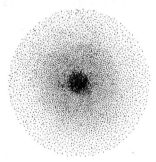

FIG. 1.1. The hydrogen atom.

greatest, and secondly that no direction is favoured over any other, in other words the electron is equally likely to exist on one side of the nucleus as the other. A description of the probability of finding an electron at any given point such as we have just discussed is called an *orbital*.

Any electron can be characterized by a set of four *quantum numbers*, the *principal* quantum number (n), the *azimuthal* quantum number (l), the *magnetic* quantum number (m) and the *spin* quantum number (s). Each of these quantum numbers can only have certain values assigned to it.

The principal quantum number can have any positive integral value.

$$n = 1, 2, 3, \ldots$$

and describes the average distance of the electron from the nucleus. This is the most important factor in determining the energy of the electron; a low value of n implies a stable low-energy electron, while a high value of n signifies an electron of high energy.

The azimuthal quantum number is determined in part by the value of n, the allowed values being

$$l = 0, 1, 2, \ldots (n-1)$$

Thus for a value of $n = 3$ only three values of l are possible, viz. 0, 1 and 2. The value of the azimuthal quantum number describes the shape of the orbital, and generally the value of l is represented by a letter rather than by a number:

Numerical value of l	0	1	2	3
Representative letter	s	p	d	f

The magnetic quantum number can assume any integral value within the limits.

$$m = 0, \pm 1, \pm 2 \cdots \pm l$$

While the value of l describes the shape of the orbital, the value of m describes the orientation of the orbital relative to the nucleus.

The spin quantum number describes the direction of spin of the electron, which may be envisaged as spinning about its own axis in either a clockwise or anti-clockwise direction. There are therefore only two possible values for s, $+\frac{1}{2}$ or $-\frac{1}{2}$.

Any electron then can be characterized by quoting values for these four quantum numbers, and in particular the energy of the electron is determined predominantly by the values of n and l. For our purposes we can assume that the energies of orbitals with the same values of n and l but different values of m and s, are the same, and only n and l values need be specified.

Another basic postulate of modern theory is that *no two electrons on the same atom can have the same quantum number values*. With this in mind we can use these definitions of the various quantum numbers to see how electrons arrange themselves around the nucleus. For a value of $n = 1$, only two electrons can exist, having the quantum numbers:

n	l	m	s
1	0	0	$+\frac{1}{2}$
1	0	0	$-\frac{1}{2}$

Similarly, for $n = 2$ we find a group of eight electrons:

n	l	m	s
2	0	0	$+\frac{1}{2}$
2	0	0	$-\frac{1}{2}$
2	1	-1	$+\frac{1}{2}$
2	1	-1	$-\frac{1}{2}$
2	1	0	$+\frac{1}{2}$
2	1	0	$-\frac{1}{2}$
2	1	$+1$	$+\frac{1}{2}$
2	1	$+1$	$-\frac{1}{2}$

For $n = 3$ there are eighteen possible combinations of quantum numbers and therefore eighteen electrons can exist with this value for the principal quantum number. The full significance of this sequence of numbers will be discussed again in Chapter 3.

We have already learned that the energy of an electron is described by the values of the principal and azimuthal quantum numbers. Figure 1.2 shows how the relative energy varies with the quantum number value, each circle in the diagram representing an orbital capable of containing two electrons. If we now accept that the electrons on any atom tend to adopt the configuration of lowest energy we can build up a picture of the electronic structure of an atom.

Hydrogen, the simplest atom with an atomic number $Z = 1$, has one electron which will be held in the $1s$ orbital, this being the orbital of lowest energy as indicated in Fig. 1.2. The electronic configuration of this atom would then be represented $1s^1$.

The atom with atomic number $Z = 2$ is helium, which has two electrons to be arranged around the nucleus. By reference to Fig. 1.2 we can see that both these electrons can be accommodated in the $1s$ orbital. Hence the electronic configuration of helium can be represented $1s^2$. At this point it can be seen that the $1s$ orbital now holds its maximum quota of electrons, and since helium is known to be chemically unreactive we can

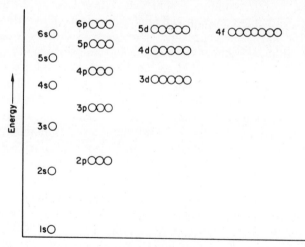

Fig. 1.2. Orbital energy levels.

say that this situation is associated with a high degree of stability.

The lithium atom has three electrons only two of which can occupy the low energy $1s$ orbital, leaving the third electron to be accommodated in the $2s$ orbital. Thus the configuration will be represented $1s^2 . 2s^1$. Again by reference to Fig. 1.2 it can be seen that eight electrons can be accommodated in the group of orbitals having a principal quantum number $n = 2$, before this group is completely filled. As is shown in Table 1.1 this situation arises with the neon atom which has the configuration $1s^2 2s^2 2p^6$. Neon, like helium, is highly inert chemically, and again we note that a high degree of stability is associated with the state where all the orbitals of a group hold their maximum number of electrons.

1.5. Chemical bonding

We have built up a picture of an isolated atom but have not as yet discussed how these atoms can combine with similar atoms or with atoms of a different element. We want to understand *why* different atoms combine to form a new molecule and also *how* the bonding is achieved. The answer to the first question is that any system will tend to adjust itself in order to attain the most stable possible state, i.e. the state of lowest energy. If it is assumed that electronic rearrangements will readily occur provided that the end result is a more stable arrangement, then the formation of bonds between different atoms is readily understood. It has already been stated in the previous section that an arrangement of electrons where a group of orbitals holds the maximum possible number of electrons is one of high stability. It is from this viewpoint, viz. that electronic rearrangements will occur to attain these configurations of high stability, that we shall attempt to account for the formation of chemical bonds.

The answer to how bonding is achieved leads to a classification of bond types under two main headings depending on what type of electron rearrangement takes place. Chemical bonds can be described either as *ionic*, where electrons are completely transferred from one atom to another, or as *co-valent*, where electrons are shared between atoms. It must be emphasized, however, that most chemical bonds are inter-mediate between these two extreme types although in practice they are usually classed as one or the other.

1.6.   Ionic bonds

A good example of the formation of an ionic bond is the formation of sodium chloride from sodium and chlorine atoms. By reference to Table 1.1 we find that the electronic configuration of the sodium atom is

$$1s^2 \quad 2s^2 2p^6 \quad 3s^1$$

We know that the configurations $1s^2$ and $2s^2 2p^6$ are stable clos-ed systems where the groups of orbitals contain their maximum number of electrons, and that the single electron in the $3s$ orbital is a fairly high energy electron. This electron is obvious-ly in a different category from the others and it is found that this electron can be easily removed from the atom leaving a *sodium ion* (Na^+) which is positively charged and whose elec-tronic structure is a known stable configuration.

Similarly we know that the electronic configuration of the chlorine atom is

$$1s^2 \quad 2s^2 2p^6 \quad 3s^2 3p^5$$

Here again we find two closed stable configurations together with an outer shell of seven electrons. It can readily be seen that if the chlorine atom could acquire an extra electron to form a negatively charged *chloride ion* (Cl^-), again a stable electronic configuration would be achieved. Thus by the transfer of one electron from a sodium atom to a chlorine atom two *ions* are produced, each with a stable arrangement of electrons, and since these ions are of opposite electrical charge they attract each other to form an ionic bond. By considering only the outer

TABLE 1.1. ELECTRONIC

Z	Element	1 s	2 s	2 p	3 s	3 p	3 d
1	H	1					
2	He	2					
3	Li	2	1				
4	Be	2	2				
5	B	2	2	1			
6	C	2	2	2			
7	N	2	2	3			
8	O	2	2	4			
9	F	2	2	5			
10	Ne	2	2	6			
11	Na	2	2	6	1		
12	Mg	2	2	6	2		
13	Al	2	2	6	2	1	
14	Si	2	2	6	2	2	
15	P	2	2	6	2	3	
16	S	2	2	6	2	4	
17	Cl	2	2	6	2	5	
18	Ar	2	2	6	2	6	
19	K	2	2	6	2	6	
20	Ca	2	2	6	2	6	
21	Sc	2	2	6	2	6	1
22	Ti	2	2	6	2	6	2
23	V	2	2	6	2	6	3
24	Cr	2	2	6	2	6	5
25	Mn	2	2	6	2	6	5
26	Fe	2	2	6	2	6	6
27	Co	2	2	6	2	6	7
28	Ni	2	2	6	2	6	8
29	Cu	2	2	6	2	6	10
30	Zn	2	2	6	2	6	10
31	Ga	2	2	6	2	6	10
32	Ge	2	2	6	2	6	10
33	As	2	2	6	2	6	10
34	Se	2	2	6	2	6	10
35	Br	2	2	6	2	6	10
36	Kr	2	2	6	2	6	10
37	Rb	2	2	6	2	6	10
38	Sr	2	2	6	2	6	10
39	Y	2	2	6	2	6	10
40	Zr	2	2	6	2	6	10
41	Nb	2	2	6	2	6	10
42	No	2	2	6	2	6	10
43	Tc	2	2	6	2	6	10

STRUCTURE OF THE ATOMS

| 4 | | | | 5 | | | | 6 | | | | 7 |
s	p	d	f	s	p	d	f	s	p	d	f	s
1												
2												
2												
2												
2												
1												
2												
2												
2												
2												
1												
2												
2	1											
2	2											
2	3											
2	4											
2	5											
2	6											
2	6			1								
2	6			2								
2	6	1		2								
2	6	2		2								
2	6	4		1								
2	6	5		1								
2	6	6		1								

TABLE 1.1.

		1	2		3		
Z	Element	s	s	p	s	p	d
44	Ru	2	2	6	2	6	10
45	Rh	2	2	6	2	6	10
46	Pd	2	2	6	2	6	10
47	Ag	2	2	6	2	6	10
48	Cd	2	2	6	2	6	10
49	In	2	2	6	2	6	10
50	Sn	2	2	6	2	6	10
51	Sb	2	2	6	2	6	10
52	Te	2	2	6	2	6	10
53	I	2	2	6	2	6	10
54	Xe	2	2	6	2	6	10
55	Cs	2	2	6	2	6	10
56	Ba	2	2	6	2	6	10
57	La	2	2	6	2	6	10
58	Ce	2	2	6	2	6	10
59	Pr	2	2	6	2	6	10
60	Nd	2	2	6	2	6	10
61	Pm	2	2	6	2	6	10
62	Sm	2	2	6	2	6	10
63	Eu	2	2	6	2	6	10
64	Gd	2	2	6	2	6	10
65	Tb	2	2	6	2	6	10
66	Dy	2	2	6	2	6	10
67	Ho	2	2	6	2	6	10
68	Er	2	2	6	2	6	10
69	Tm	2	2	6	2	6	10
70	Yb	2	2	6	2	6	10
71	Lu	2	2	6	2	6	10
72	Hf	2	2	6	2	6	10
73	Ta	2	2	6	2	6	10
74	W	2	2	6	2	6	10
75	Re	2	2	6	2	6	10
76	Os	2	2	6	2	6	10
77	Ir	2	2	6	2	6	10
78	Pt	2	2	6	2	6	10
79	Au	2	2	6	2	6	10
80	Hg	2	2	6	2	6	10
81	Tl	2	2	6	2	6	10
82	Pb	2	2	6	2	6	10
83	Bi	2	2	6	2	6	10
84	Po	2	2	6	2	6	10
85	At	2	2	6	2	6	10
86	Rn	2	2	6	2	6	10

(cont.)

	4				5				6				7
s	p	d	f	s	p	d	f	s	p	d	f	s
2	6	7		1								
2	6	8		1								
2	6	10										
2	6	10		1								
2	6	10		2								
2	6	10		2	1							
2	6	10		2	2							
2	6	10		2	3							
2	6	10		2	4							
2	6	10		2	5							
2	6	10		2	6							
2	6	10		2	6			1				
2	6	10		2	6			2				
2	6	10		2	6	1		2				
2	6	10	2	2	6			2				
2	6	10	3	2	6			2				
2	6	10	4	2	6			2				
2	6	10	5	2	6			2				
2	6	10	6	2	6			2				
2	6	10	7	2	6			2				
2	6	10	7	2	6	1		2				
2	6	10	9	2	6			2				
2	6	10	10	2	6			2				
2	6	10	11	2	6			2				
2	6	10	12	2	6			2				
2	6	10	13	2	6			2				
2	6	10	14	2	6			2				
2	6	10	14	2	6	1		2				
2	6	10	14	2	6	2		2				
2	6	10	14	2	6	3		2				
2	6	10	14	2	6	4		2				
2	6	10	14	2	6	5		2				
2	6	10	14	2	6	6		2				
2	6	10	14	2	6	9						
2	6	10	14	2	6	9		1				
2	6	10	14	2	6	10		1				
2	6	10	14	2	6	10		2				
2	6	10	14	2	6	10		2	1			
2	6	10	14	2	6	10		2	2			
2	6	10	14	2	6	10		2	3			
2	6	10	14	2	6	10		2	4			
2	6	10	14	2	6	10		2	5			
2	6	10	14	2	6	10		2	6			

Table 1.1.

Z	Element	1 s	2 s	p	3 s	p	d
87	Fr	2	2	6	2	6	10
88	Ra	2	2	6	2	6	10
89	Ac	2	2	6	2	6	10
90	Th	2	2	6	2	6	10
91	Pa	2	2	6	2	6	10
92	U	2	2	6	2	6	10
93	Np	2	2	6	2	6	10
94	Pu	2	2	6	2	6	10
95	Am	2	2	6	2	6	10
96	Cm	2	2	6	2	6	10
97	Bk	2	2	6	2	6	10
98	Cf	2	2	6	2	6	10
99	Es	2	2	6	2	6	10
100	Fm	2	2	6	2	6	10
101	Md	2	2	6	2	6	10
102	No	2	2	6	2	6	10
103	Lw	2	2	6	2	6	10

electrons, the bond formation may be represented

$$\text{Na}° + \cdot \ddot{\underset{\cdot\cdot}{\text{Cl}}}\text{:} \rightarrow \text{Na}^+ \text{:}\ddot{\underset{\cdot\cdot}{\text{Cl}}}\text{:}^-$$

It is important to remember that in the formation of a bond of this type there must be a balance between the electrons lost and the electrons gained. If we consider the formation of another compound, calcium chloride, we find a slightly different situation, Calcium has the configuration:

$$1s^2\,2s^2 2p^6\,3s^2 3p^6\,4s^2$$

Here *two* electrons must be transferred from the atom in order to attain a stable configuration, hence *two* chlorine atoms must be involved to maintain the electron balance.

$$\text{:}\ddot{\underset{\cdot\cdot}{\text{Cl}}}\cdot + \overset{\circ\circ}{\text{Ca}} + \cdot\ddot{\underset{\cdot\cdot}{\text{Cl}}}\text{:} \rightarrow \text{:}\ddot{\underset{\cdot\cdot}{\text{Cl}}}\text{:}^-\ \text{Ca}^{2+}\ \text{:}\ddot{\underset{\cdot\cdot}{\text{Cl}}}\text{:}$$

The term valency is used to describe the ability of atoms to bond together. From the above examples sodium is said to have a valency of one, and calcium a valency of two.

(cont.)

4 s	p	d	f	5 s	p	d	f	6 s	p	d	f	7 s
2	6	10	14	2	6	10		2	6			1
2	6	10	14	2	6	10		2	6			2
2	6	10	14	2	6	10		2	6	1		2
2	6	10	14	2	6	10		2	6	2		2
2	6	10	14	2	6	10	2	2	6	1		2
2	6	10	14	2	6	10	3	2	6	1		2
2	6	10	14	2	6	10	4	2	6	1		2
2	6	10	14	2	6	10	6	2	6			2
2	6	10	14	2	6	10	7	2	6			2
2	6	10	14	2	6	10	7	2	6	1		2
2	6	10	14	2	6	10	8	2	6	1		2
2	6	10	14	2	6	10	10	2	6			2
2	6	10	14	2	6	10	11	2	6			2
2	6	10	14	2	6	10	12	2	6			2
2	6	10	14	2	6	10	13	2	6			2
2	6	10	14	2	6	10	13	2	6	1		2
2	6	10	14	2	6	10	14	2	6	1		2

1.7. Covalent bonds

It can easily be seen that only a limited number of elements will be capable of forming bonds by the total transfer of electrons. Since negatively charged electrons are attracted to the positively charged nucleus by electrostatic forces, there will obviously be a limit to the number of electrons which can be removed from any particular atom, because the removal of an electron increases the residual positive charge by which the others are retained. In general it is energetically not feasible to transfer more than two electrons. For this reason the majority of compounds are formed by *covalent* bonding, i.e. by the *sharing* of electrons.

The simplest example of this type of bonding is the formation of a hydrogen molecule from two atoms of hydrogen. We have already learned that the hydrogen atom has only a single 1s electron and that the 1s orbital is capable of holding two electrons. Although it would be possible to describe bond formation between two hydrogen atoms formally as an ionic bond:

$$H^\bullet + H^\circ \rightarrow H^+{}_\circ^\circ H^-$$

it seems unreasonable to suppose that one hydrogen atom should be capable of abstracting an electron from another exactly similar hydrogen atom. Instead, the bond is described as the sharing of the two electrons between the two atoms:

$$H^\bullet + H^\circ \rightarrow H{}_\circ^\circ H$$

Just as the hydrogen atom can be represented as in Fig. 1.1 the hydrogen molecule can be pictured as in Fig. 1.3 where the electron clouds of the separate hydrogen atoms have merged to produce a region of high electron density between the two atoms, thus forming a covalent bond.

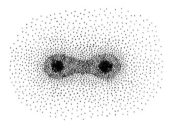

Fig. 1.3. The hydrogen molecule.

A slightly more complicated example is the formation of the hydrogen chloride molecule. Here the hydrogen atom has one electron and the chlorine atom has seven outer electrons, hence covalent bonding can occur to give the hydrogen atom its full complement of two electrons and the chlorine atom a stable arrangement of eight electrons:

$$H^\bullet + {}_\circ\overset{\circ\circ}{\underset{\circ\circ}{C}}l_\circ^\circ \rightarrow H{}_\circ^\bullet\overset{\circ\circ}{\underset{\circ\circ}{C}}l_\circ^\circ$$

Although the H—Cl bond is covalent it differs in one important respect from the covalent H—H bond. In the hydrogen molecule the pair of electrons forming the bond is shared equally between the two atoms, whereas in the hydrogen chloride molecule this is not the case. Because of differences in internal atomic structure atoms of different elements vary in their capacity to attract electrons. An element that tends to acquire rather than lose electrons in its reactions is said to be *electronegative*, while an element: which tends to lose electrons is described as *electropositive*. In the above example the chlorine atom is more electronegative than hydrogen, thus the electron pair forming the bond will have a greater probability of being found near the chlorine atom, which will therefore tend to be negatively charged with respect to the hydrogen. Such a bond is described as *polar*, and will be expected to influence the properties of the material.

A molecule can obviously contain more than two atoms. For example, in a water molecule the oxygen atom, which has six outer electrons, forms two covalent bonds with hydrogen atoms in order to attain a stable configuration:

$$H^{\bullet} + {}^{\circ}\overset{\circ\circ}{\underset{\circ\circ}{O}}{}^{\circ} + H^{\bullet} \rightarrow H \overset{\circ\circ}{\underset{\circ\circ}{O}} H$$

As was the case in ionic bonding the valency of a particular atom is determined by its electronic structure, and in the examples we have considered the valency of hydrogen and of chlorine is one, while that of oxygen is two.

If there is an insufficient number of electrons to attain a stable configuration by the sharing of one electron pair, it is possible for two atoms to share two pairs to form a *double* bond. An example of this is the carbon dioxide molecule, where the carbon atom has four valency electrons and the oxygen atoms each have six:

$$^{\circ}\overset{\circ\circ}{\underset{\circ\circ}{O}}{}^{\circ} + \overset{\bullet\bullet}{\underset{\bullet\bullet}{C}} + {}^{\circ}\overset{\circ\circ}{\underset{\circ\circ}{O}}{}^{\circ} \rightarrow \overset{\circ\circ}{\underset{\circ\circ}{O}} \overset{\bullet}{\underset{\bullet}{C}} \overset{\circ\circ}{\underset{\circ\circ}{O}}$$

Similarly the formation of a *triple* bond may occur when three electron pairs must be shared between two atoms to attain a stable configuration. This occurs, for example, in the nitrogen molecule, where each nitrogen atom possesses only five valency electrons.

$$:\overset{\circ}{\underset{\circ}{N}}: + \cdot\overset{\bullet}{\underset{\bullet}{N}}: \rightarrow :N::N:$$

In all examples considered up to this point a covalent bond has been formed by the sharing of a pair of electrons, one electron being donated by each atom. There are occasions where the two electrons required for bond formation are donated by only one of the atoms concerned. An example of this is the formation of the ammonium ion (NH_4^+) which may be regarded as the combination of a hydrogen ion (H^+) with a molecule of ammonia:

$$
\begin{array}{ccc}
\text{H} & & \text{H} \\[4pt]
\text{H} : \overset{\cdot\cdot}{\underset{\cdot\cdot}{N}} : + \quad \text{H}^+ \rightarrow & \left[\text{H} \quad : \overset{\cdot\cdot}{\underset{\cdot\cdot}{N}} : {}^{\text{J}}\text{H} \right]^+ \\[4pt]
\text{H} & & \text{H}
\end{array}
$$

Here the ammonia molecule contains three covalent $N-H$ bonds, leaving a *lone pair* of electrons not involved in bond formation. This lone pair of electrons is available for sharing with the hydrogen ion to form a bond which is usually described as a *co-ordinate covalent bond*.

1.8. Hybridization

Since a large part of this book deals with the chemistry of carbon compounds, it is worth while at this stage to study in rather more detail the bonding involved in such compounds.

The electronic configuration of carbon is

$$1s^2 \qquad 2s^2 2p^2$$

i.e. there exists a stable closed system of two electrons, leaving four valency electrons.

In Section 1.4 we learned that the shape of an orbital was described by the value of the azimuthal quantum number and that an s orbital was spherically symmetrical as shown in Fig. 1.1. There are three p orbitals each of which can hold two electrons, and these dumb-bell-shaped orbitals are arranged mutually at right angles to one another as is shown in Fig. 1.4.

We now have a situation where a single carbon atom is pictured as having two valency electrons in a spherical s orbital, and two in two of the three p orbitals. On the other hand,

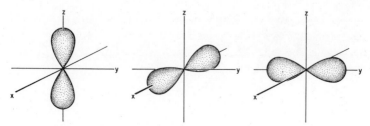

FIG. 1.4. p-orbitals.

since it is found from experiment that when carbon forms a compound such as methane the four bonds are equivalent, we must modify our picture to obtain four equivalent orbitals around the carbon atom. This is achieved by postulating a process of *hybridization* where by 'blending' one s orbital and three p orbitals we obtain four sp^3 orbitals directed tetrahedrally (i.e. mutually at an angle of $109°28'$) as shown in Fig. 1.5(a). Since the formation of a covalent bond can be pictured as the overlapping or merging of the electron orbitals, we would picture the molecule of methane as in Fig. 1.5(b) with the central carbon atom surrounded tetrahedrally by the four hydrogen atoms.

When a carbon–carbon double bond is formed, as is the case in ethylene ($CH_2{=}CH_2$), a different type of hybridization occurs. Here one s orbital and two p orbitals are blended to give three equivalent sp^2 hybrids, mutually at an angle of $120°$, leaving one p orbital unchanged. (Fig. 1.6(a))

Again bond formation has involved the overlapping of orbitals to give a flat molecule where each carbon atom forms three normal covalent bonds (σ bonds) and a second carbon–carbon bond (π bond) is formed by the merging of the unhybridized p orbitals above and below the plane of the molecule

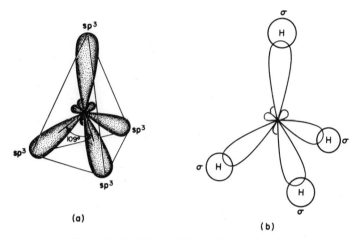

FIG. 1.5. *sp³*-orbitals and the methane molecule.

in Fig. 1.6(b). It can be seen that this second carbon–carbon bond is much more diffuse than the others in the molecule and is found to be more susceptible to chemical attack as will be explained in Chapter 20.

An analogous situation is found in acetylene (CH≡CH) where a triple bond is formed. In this case only one *p* orbital is combined with the *s* orbital to give two *sp* orbitals and the remaining two *p* orbitals are left unchanged as shown in Fig. 1.7(a). If bond formation is again pictured as the overlapping

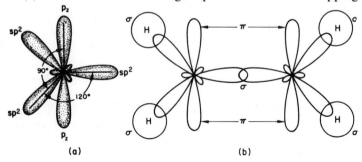

FIG. 1.6. *sp²*-orbitals and the ethylene molecule.

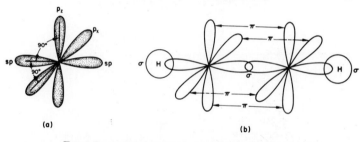

Fig. 1.7. *sp*-orbitals and the acetylene molecule.

of orbitals it can be seen that each carbon atom will form two normal σ bonds, while at the same time two diffuse π bonds are formed between the carbon atoms as illustrated in Fig. 1.7(b).

If we now consider a molecule containing *conjugated* double bonds, i.e. alternate single and double bonds, we find a further complication. A molecule of this type is *benzene* whose formula may be written

$$
\begin{array}{ccc}
 & H & \\
H & C & H \\
\diagdown & \| & \diagup \\
C & & C \\
| & & \| \\
C & & C \\
\diagup & & \diagdown \\
H & C & H \\
 & | & \\
 & H & \\
\end{array}
$$

By analogy with ethylene we might expect that sp^2 hybridization would occur giving a planar cyclic molecule containing three well defined double bonds. This description does not explain the observed chemistry of benzene, as you will learn in Chapter 21.

We must picture the molecule as in Fig. 1.8. Instead of having three separate electron clouds forming three well-defined, separate double bonds, we allow those isolated clouds to merge to form a continuous electron cloud above and below the plane

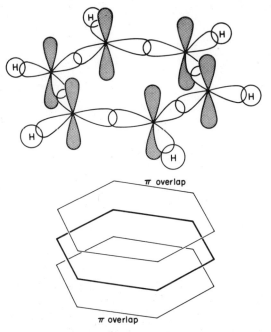

FIG. 1.8. *sp²*-orbitals and the benzene molecule.

of the molecule. This implies that the electrons forming the double bonds are not localized between any two carbon atoms but are free to move around the molecule, imparting a certain amount of double bond character to each carbon–carbon bond. The electrons are said to be *delocalized*. Any molecule where delocalization of electrons is possible is found to be far more stable than would otherwise be expected.

1.9. The hydrogen bond

It is convenient to introduce at this point an unusual feature of certain compounds containing hydrogen. It is found that where molecules contain hydrogen bonded to a highly electronegative element, e.g. −OH, −NH, −FH, unexpected forces of attraction exist which may considerably affect the properties of the material. If we compare the boiling points of the hydrogen halides (Table 1.2) we see that for the latter members of the group the boiling points increase with increasing molecular weight. Hydrogen fluoride, however, has an abnormally high boiling point, showing that the molecules of this material must be held together in some unusual way.

TABLE 1.2. BOILING POINTS OF THE
HYDROGEN HALIDES

Compound	HF	HCl	HBr	HI
Boiling point (°C)	19·4	−83·1	−68·7	−36·7

Since a hydrogen atom has only a single $1s$ electron it can form only one normal covalent bond. Nevertheless, it can at the same time form a special type of bond called the *hydrogen bond*, which is much weaker, usually by a factor of about twenty, than the normal covalent type. The bond may be envisaged as being mainly electrostatic in character, the attraction being between the electropositive hydrogen and a lone pair of electrons on the electronegative element involved.

The abnormally high boiling point of hydrogen fluoride can now be understood in terms of the individual molecules being linked together by hydrogen bond 'bridges' which must be broken before vaporization can occur.

Since the elements chlorine, bromine and iodine are much less electronegative than fluorine, hydrogen bonding is negligible in the hydrides of these elements.

Hydrogen bonding involving oxygen atoms is of considerable importance in organic chemistry and drastically modifies the physical properties of many organic compounds. The lower

TABLE 1.3. PHYSICAL PROPERTIES OF SOME
LOWER ALCOHOLS AND HYDROCARBONS

Compound	Molecular weight	Melting point (°C)	Boiling point (°C)
CH_3OH	32	−97·8	64·7
CH_3CH_3	30	−172	−89·0
CH_3CH_2OH	46	−117·6	78·5
$CH_3CH_2CH_3$	44	−189·9	−42·2

alcohols, for example, have properties very different from closely related compounds of similar molecular weight (Table 1.3) because of the effect of hydrogen bond bridging between molecules:

$$O-H\text{--}O-H\text{--}O-H\text{--}O-H$$
$$\quad R \qquad R \qquad R \qquad R$$

CHAPTER 2

Radioactivity

WE HAVE already learned that for the chemist it is the *electronic structure* of the atoms which governs chemical behaviour. In normal chemical reactions the atomic nuclei remain unchanged and the formation of new chemical bonds is ascribed to a rearrangement of electrons in the system. There are certain types of reaction, however, where changes take place within the nucleus itself. These reactions are accompanied by very large energy changes, and elements which undergo such reactions spontaneously are described as *radioactive*.

Natural radioactivity was first discovered in 1896 by Becquerel, who observed that uranium salts emitted some kind of high energy radiation which had the effect of blackening a photographic plate. Soon after this, Pierre and Marie Curie isolated a new radioactive element, radium, and established that the radiation was an atomic phenomenon, characteristic of the element involved.

2.1. Nuclear stability

We learned in Chapter 1 that the nucleus of an atom is composed essentially of protons and neutrons, and that these particles are held together in a very small volume by strong attractive forces. At the same time it is clear that a number of positively charged protons held in a small volume must repel one another, and this repulsive force must increase as the number of protons increases. It is found that as the number of protons in a nucleus increases, so the number of neutron present

25

increases to overcome this repulsive force between the protons; that is, as the atomic number increases, so does the neutron: proton ($n:p$) ratio. This is shown graphically in Fig. 2.1, where the number of neutrons is plotted against the number of protons for naturally occurring stable nuclei.

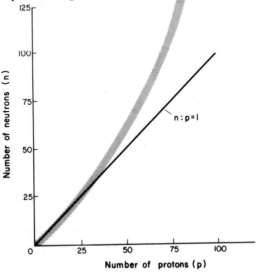

Fig. 2.1. $n:p$ ratios in stable nuclei.

We see that the $n:p$ ratio must fall within a certain range for the nucleus to be stable, and that this ratio increases as the atomic number increases. Furthermore, we find that there is no stable ratio for elements having an atomic number greater than 83. If, therefore, a nucleus has a structure outside this stable $n:p$ ratio range it will tend to undergo a spontaneous change in order to reach a more stable state.

2.2.   Types of radioactive decay

In terms of the graph given in Fig. 2.1 we see that three types of unstable nucleus can exist:

(i) a nucleus with too high a $n{:}p$ ratio,

(ii) a nucleus with too low a $n{:}p$ ratio,

(iii) a nucleus with an atomic number greater than 83.

A nucleus of the first type, where the ratio of neutrons to protons is too high, can reduce this ratio by emitting *beta radiation*. Beta radiation consists simply of a stream of electrons and is usually described either by the symbol

$$\beta^- \text{ or } _{-1}e^0$$

$$\text{e.g. } _6C^{14} \rightarrow _7N^{14} + _{-1}e^0$$

Note that the usual way of writing such an equation involves describing each species present in the form $_aX^b$, where X is the usual symbol for the element involved, the subscript a is the atomic number and the superscript b the mass number. We see that the above example involves the isotope of carbon containing six protons and eight neutrons. Reference to Fig. 2.1 shows that this $n{:}p$ ratio is too high for stability, therefore β-emission occurs. It is formally assumed that one neutron in the unstable nucleus breaks up to yield a proton and an electron, which is then emitted:

$$\text{neutron} \rightleftharpoons \text{proton} + \text{electron}$$

Since the mass of the emitted electron is negligible, the overall effect of β-radiation is that the mass number remains unchanged while the atomic number increases its value by one.

Another possible method of reducing the $n:p$ ratio is simply by the emission of unwanted neutrons from the unstable nucleus. While this reaction can occur, it is extremely uncommon.

Nuclei of the second type, where the $n:p$ ratio is too low, undergo nuclear reactions which decrease their nuclear charge. Thus whereas nuclei with a high $n:p$ ratio emit electrons, nuclei with low $n:p$ ratios emit *positrons* (particles having the same mass as an electron, but carrying a positive charge). A proton in the unstable nucleus is regarded as breaking up to form a neutron and a positron ($_1e^0$) which is then emitted.

$$proton \rightleftharpoons neutron + positron$$

e.g. $_{10}Ne^{19} \rightarrow {}_9F^{19} + {}_1e^0$

In this example the unstable neon nucleus undergoes positron emission. Since the mass of a positron is negligible the mass number of the nucleus is unchanged, but the atomic number is decreased by one unit.

Another method of increasing the $n:p$ ratio is by *electron capture*. Here one of the orbital electrons is captured by the nucleus thus converting one of the nuclear protons to a neutron and thereby increasing the $n:p$ ratio to a more stable value,

e.g. $_{18}A^{37} \xrightarrow[\text{capture}]{\text{electron}} {}_{17}Cl^{37}$

The third type of unstable nucleus, where the atomic number is greater than 83, usually achieves stability by emitting *alpha particles* (written α or $_2He^4$), which are *dipositive helium atoms*,

e.g. $_{92}U^{238} \rightarrow {}_{90}Th^{234} + {}_2He^4$

Here we start with a uranium isotope which is unstable because of its high atomic number and consequently undergoes α-emission to form the element thorium. This isotope is still unstable and emits β-particles to form the element protactinium.

$$_{90}Th^{234} \rightarrow {}_{91}Pa^{234} + {}_{-1}e^0$$

This nucleus is still unstable and a series of nuclear reactions takes place, involving the emission of both α- and β-particles until eventually the stable lead isotope $_{82}Pb^{206}$ is obtained.

One further type of radiation must be described, namely *gamma-rays* (γ-rays). These do not consist of particles and are neither positively nor negatively charged, but are high energy electromagnetic waves, similar to X-rays, and are often emitted when a nucleus undergoes a disintegration reaction.

2.3. Rate of radioactive decay

The rate at which the above reactions occur is independent of external influences such as temperature or pressure, and depends on only two factors, viz. the type of nucleus involved and the number of atoms present. In other words, for any given radioactive element a definite fraction of the nuclei present will disintegrate per unit time, thus the greater the number of atoms present the more disintegrations will occur. The rate of decay of any species is described by the term *half-life* which is the time required for half the number of atoms present to disintegrate.

The half-life is a measure of the stability of any given element, the shorter the half-life the more unstable the nucleus. Since the stabilities of different radioactive elements vary enormously, the half-life can vary from millions of years to only a fraction of a second.

2.4. Radioactive dating

A knowledge of the rate of decay of radioactive isotopes makes it possible for us to determine the age of materials containing these isotopes. For example, a radioactive isotope of carbon with a half-life of 5700 years is present in the atmosphere in the form of carbon dioxide. This radioactive carbon is assimilated by all living matter which therefore contains a known amount of radioactive carbon. At death the absorption of radioactive carbon ceases and the amount present in the material steadily decreases due to radioactive decay. By measuring the activity remaining, the age of the material can be calculated.

The radiocarbon method is suitable only for the determination of ages of thousands of years. The ages of rocks millions of years old can, however, be determined by a similar method based on the decay of uranium to stable lead isotopes. Since the half-life of uranium is $4 \cdot 5 \times 10^9$ years, this method can be used to date rocks 500 million years old, although the errors involved may be quite considerable.

2.5. Artificial radioactivity

We have discussed how some naturally occurring elements may have nuclei which are unstable and which therefore spontaneously disintegrate to form a stable nucleus, at the same time emitting some sort of radiation. We must now consider what would happen if these reactions were reversible; i.e. the effect of bombarding a stable nucleus with high energy particles.

Such a situation was first observed by Rutherford in 1919 when he irradiated nitrogen with high-energy α-particles and found that an oxygen isotope was produced along with protons. The reaction may be represented

$$_7N^{14} + _2He^4 \rightarrow _8O^{17} + _1H^1$$

Here the nitrogen nucleus has absorbed an α-particle to form an unstable intermediate which then disintegrates to form oxygen. Other positively charged projectiles used are protons ($_1H^1$) and deuterons ($_1H^2$).

Although each of these projectiles can successfully induce nuclear reactions they all suffer from one disadvantage, viz. they are positively charged. Since the experiment involves the penetration of a positively charged nucleus by one of the bombarding particles, it can easily be seen that the particles will need to possess a very large amount of energy in order to overcome the repulsive force of the nucleus. The necessary energy can be achieved by means of *particle accelerators* where the particles can be made to move with very high velocities by subjecting them to large potential differences, but the process is fairly costly.

For this reason, the most useful projectile is the neutron which is not repelled by the nucleus and therefore does not require a high velocity, but can be absorbed at low speeds. If we bombard a light element such as aluminium with neutrons a variety of reactions can take place. For example, if high-velocity neutrons are used we get reactions similar to the bombardment of nitrogen with α-particles.

$$_{13}Al^{27} + _{0}n^{1} \rightarrow _{11}Na^{24} + _{2}He^{4}$$
$$_{13}Al^{27} + _{0}n^{1} \rightarrow _{12}Mg^{27} + _{1}H^{1}$$

In each case new particles are ejected from the nucleus and a different element is produced. Both the sodium and magnesium isotopes are in fact unstable and undergo β-decay.

If instead of fast neutrons, slow neutrons are used, a different type of reaction results.

$$_{13}Al^{27} + _{0}n^{1} \rightarrow _{13}Al^{28}$$

$$_{12}Mg^{27} + _{1}H^{1} \qquad _{13}Al^{28} + \gamma \qquad _{13}Al^{26} + 2_{0}n^{1}$$

Here the bombarding neutron is *captured* by the nucleus to form a new isotope which may eventually break down in a variety of ways. Note that in one case we obtain an extra neutron, a process known as *spallation*.

2.6. Energy changes

We learned at the beginning of this chapter that nuclear reactions involved very large amounts of energy, but we have not yet discussed the source of this energy. Normal chemical reactions do not usually involve energy changes much greater than 100 kcal mole^{-1}, whereas the energy changes involved in nuclear transformations may be a million times greater than this.

The source of this vast amount of energy is found in the Einstein relationship, $E = mc^2$, where E is the energy, m the mass and c the velocity of light. The law of conservation of mass, which states that mass is neither created nor destroyed, can be safely applied to normal chemical reactions where energy changes are small. If however we try to apply this law to nuclear changes, we find that the 'law' appears to be invalid.

Thus we know that a deuteron (deuterium nucleus) is composed of one neutron and one proton, and we might expect that the mass of a deuteron would equal the mass of a proton (1·00783 a.m.u.) plus that of a neutron (1·00867 a.m.u.), viz. 2·01650 a.m.u. However, the experimental value for the mass of a deuteron is 2·01410 a.m.u., implying that there is a mass loss of 0·00240 units. This mass loss is explained by the fact that the Einstein relationship shows mass and energy to be equivalent.

In a nuclear reaction we should think then not in terms of conservation of mass, but of mass-energy, i.e.

$$neutron + proton = deuteron + energy.$$

That the amount of energy involved is very large can easily be seen by calculating the energy equivalent of a mass of 1 gram:

$$E = mc^2$$
$$= 1 \times (3 \times 10^{10})^2$$
$$= 9 \times 10^{20} \text{ ergs}$$
$$= 2·15 \times 10^{10} \text{ kcal.}$$

2.7. Nuclear fission

Although all the nuclear reactions we have discussed so far are potentially sources of large amounts of energy, in practice these reactions can only be carried out on a fairly small scale. However, in 1939 a different type of nuclear reaction was discovered. It was found that when uranium was bombarded by neutrons the large nucleus split into two fragments of approximately equal mass, at the same time liberating large amounts of energy and several neutrons. These extra neutrons can in turn cause the fission of other nuclei thus generating a *chain reaction*, provided that none of the neutrons are lost by side reactions.

The rate at which neutrons escape from a sample of uranium depends on the amount of material present and its shape. If the effective surface area is large the neutrons can readily escape, but if the size of the sample is increased a point is reached where the rate of generation of neutrons exceeds the rate of loss and the chain reaction develops. There is therefore a *critical mass*, below which only a slow disintegration will occur, but above which an explosive chain reaction develops.

Obviously if the vast amount of energy released in such a reaction is to be harnessed the chain reaction must be controlled in some way. This can be achieved in a nuclear reactor where the fissionable material is surrounded by graphite or heavy water which acts as a *moderator*. This slows down the fast neutrons released by the disintegrating nuclei and increases the chance of their capture by another nucleus. The rate of reaction is controlled by means of rods of neutron absorbing materials (e.g. cadmium, boron) which can be lowered into the reaction when necessary. The energy released appears mainly as heat which can then be used to generate electricity by conventional methods.

2.8. Nuclear fusion

Although a great amount of energy is released from a nuclear fission reaction, the mass changes involved are relatively small. It was realized that if *nuclear fusion* could be achieved, even greater amounts of energy would be released. Nuclear fusion involves the formation of a single nucleus from two smaller nuclei. For example, it is believed that the main source of the Sun's energy arises from the conversion of hydrogen to helium by the following reaction sequence:

$$_1H^1 + _1H^1 \rightarrow _1H^2 + _1e^0$$
$$_1H^2 + _1H^1 \rightarrow _2He^3 + \gamma$$
$$_2He^3 + _2He^3 \rightarrow _2He^4 + _2H^1 + _1e^0$$

The mass difference between helium and the four hydrogen atoms is $0 \cdot 2759$ a.m.u., and this appears as an equivalent amount of energy as explained in Section 2.6.

The main disadvantage of this type of reaction is that since it involves the fusion of two positively charged nuclei very high energies are required to bring the reaction about. While it is easy to appreciate that in the Sun, where temperatures are of the order of millions of degrees, such a reaction can proceed, it is obviously difficult to carry out fusion reactions at will. They can be achieved, however, by using the energy released in a nuclear fission reaction to produce the required high temperatures.

2.9. Miscellaneous applications

Although it is only relatively recently that radioisotopes have become readily available their special properties have found application in many different fields, only a few of which can be mentioned here.

The most spectacular application of harnessing the energy evolved from nuclear reactions has already been discussed in the previous sections. However, the energy can be used in other ways, e.g. medicinally in the selective destruction of certain types of cells, or for the sterilization of foodstuffs. Another interesting application is in the manufacture of luminous paints, where a simple phosphor is combined with a very small amount of radioactive material to give a constant supply of energy in the form of light.

One of the most useful characteristics of radioactive isotopes is that their radiation can be detected and measured so easily. This leads to the possibility of following the course of a process in a way that would have been impossible previously. For example, the rate of wear of bearings in an engine can be easily observed by rendering the bearing radioactive and then periodically checking the activity of the engine oil.

Radioactive isotopes have been widely used in analytical chemistry and as 'tracers' to help elucidate the mechanisms of chemical reactions.

CHAPTER 3

The Periodic Table

3.1. The periodic table

Very early in the history of chemistry it was realized that the ever-increasing knowledge of chemical behaviour must be systematized. Many attempted this task, but it was not until the latter half of the nineteenth century that much success was achieved when Mendeleeff and Meyer independently classified the known elements in order of increasing atomic weight and observed a significant pattern of chemical behaviour. It is a refinement of this idea that leads to the modern version of the periodic classification of elements.

Even now new attempts are being made to present the periodic table in a more convenient form, but by far the most usual way is that represented in Fig. 3.1. Here the elements are arranged in order of increasing atomic number and are arranged horizontally in *periods* and vertically in *groups*.

3.2. Electronic structure

While we must remember that the periodic classification of elements was achieved by considering the chemical properties of the elements, it is perhaps more satisfactory to consider how the above classification is a direct consequence of the electronic structure of atoms which we discussed in Chapter 1.

The first period contains only two elements, hydrogen and helium, whose electronic configurations are $1s^1$ and $1s^2$ respectively. You will recall that because of the rules governing the energies of electrons it is impossible to have more than two

1 H																	2 He
3 Li	4 Be											5 B	6 C	7 N	8 O	9 F	10 Ne
11 Na	12 Mg											13 Al	14 Si	15 P	16 S	17 Cl	18 Ar
19 K	20 Ca	21 Sc	22 Ti	23 V	24 Cr	25 Mn	26 Fe	27 Co	28 Ni	29 Cu	30 Zn	31 Ga	32 Ge	33 As	34 Se	35 Br	36 Kr
37 Rb	38 Sr	39 Y	40 Zr	41 Nb	42 Mo	43 Tc	44 Ru	45 Rh	46 Pd	47 Ag	48 Cd	49 In	50 Sn	51 Sb	52 Te	53 I	54 Xe
55 Cs	56 Ba	57 *La	72 Hf	73 Ta	74 W	75 Re	76 Os	77 Ir	78 Pt	79 Au	80 Hg	81 Tl	82 Pb	83 Bi	84 Po	85 At	86 Rn
87 Fr	88 Ra	89 †Ac															

*Lanthanide series		58 Ce	59 Pr	60 Nd	61 Pm	62 Sm	63 Eu	64 Gd	65 Tb	66 Dy	67 Ho	68 Er	69 Tm	70 Yb	71 Lu
†Actinide series		90 Th	91 Pa	92 U	93 Np	94 Pu	95 Am	96 Cm	97 Bk	98 Cf	99 Es	100 Fm	101 Md	102 No	103 Lw

FIG. 3.1. Periodic table of the elements.

electrons with a principal quantum number of unity. Hence the first period corresponds to filling up the $1s$ atomic orbital.

The second period contains eight elements from lithium ($Z = 3$) to neon ($Z = 10$). The outer electrons, i.e. the electrons which principally govern chemical behaviour, are characterized by having a principal quantum number equal to two. Thus, as we proceed from the element lithium, electrons are progressively added to the $2s$ and $2p$ orbitals, and as was explained in Chapter 1, these orbitals are completely filled when eight electrons have been added, i.e. at the element neon.

The third period corresponds very closely to the second. Again the period contains eight elements which in this case correspond to the progressive addition of electrons to the $3s$ and $3p$ orbitals, these being completely filled at the element argon.

Consideration of Fig. 1.2 shows that at the beginning of the fourth period two possibilities exist — electrons may enter the $4s$ orbital or the $3d$ orbitals, since these are approximately of equal energy. In fact, we find that for potassium and calcium the electrons enter the $4s$ orbital. Commencing at the element scandium, however, electrons begin to fill the available $3d$ orbitals giving rise to the *first transition series* of elements. The addition of electrons to the $3d$ orbitals continues up to the element zinc where the $3d$ orbitals are completely filled. The remaining six elements of this period, from gallium to krypton, correspond to the completion of the $4p$ orbitals.

The fifth period containing the eighteen elements from rubidium to xenon is analogous to the preceding period, the orbitals concerned being the $5s$, $4d$ and $5p$. Again we find a series of eight elements similar to the early periods interrupted by a *second transition series* of ten elements.

The sixth and seventh periods are built up in a similar fashion, except that in these periods an *inner transition series* of fourteen elements is inserted after the first element of the normal transition series. These groups of fourteen elements correspond to the filling of f-orbitals, but an account of the chemistry of these elements is outside the scope of this text.

3.3. Atomic radii

We have seen in the previous section that the electronic structure of the elements is a periodic function of their atomic numbers, i.e. a pattern exists which repeats itself in a more or less regular fashion. It would seem reasonable to assume that other atomic properties may exhibit periodicity.

A particularly important atomic property is that of size. The *atomic radius* cannot be defined exactly because of the diffuse nature of the electronic structure. Also, the atomic radius varies slightly depending on the precise nature of the bonds holding it to its neighbours. We can, however, define the *covalent atomic radius* as being half the distance between the nuclei of atoms in a suitable molecule and this may be used to compare the relative sizes of atoms.

In Fig. 3.2 the atomic radii are plotted against atomic number and we can readily see that atomic size is a periodic function of atomic number. Two main features may be distinguished. First, if we consider how the atomic radius varies along any given period, we see that the radii progressively decrease as we move along the period only to rise sharply at the last member of that particular period. This is so because, although throughout the period extra electrons are being added, at the same time the nuclear charge is being increased. The overall effect is that the electrons are attracted more strongly to the nucleus, the greater the nuclear charge. Second, we can consider the variation of atomic radii within a given group. In this case we see a different trend, i.e. the atomic radius in-

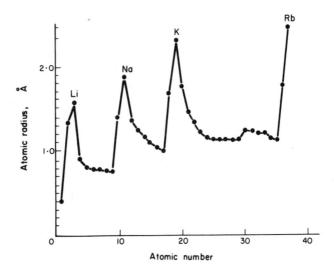

Fig. 3.2. Atomic radii as a function of atomic number.

creases with atomic number. The reason for this is that the increase in nuclear charge is more than cancelled out by the 'screening' effect of the extra shells of electrons added. The outer electrons are therefore less firmly bound to the nucleus as we descend any group in the periodic table.

3.4. Electronegativity

A useful concept for explaining some aspects of the behaviour of molecules is that of *electronegativity*, which is defined as the power of an atom to attract electrons to itself. Since the bonding in any molecule depends on the behaviour of the electrons, it is important to have some knowledge of the probable effect of any particular atom on the other electrons in the molecule.

Numerical values for the electronegativities of the elements have been obtained and Fig. 3.3 shows that again a periodic variation occurs. We see that the values range from 0·7 for

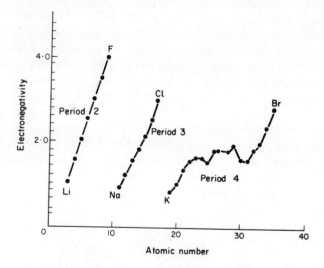

Fig. 3.3. Electronegativity as a function of atomic number.

caesium and francium, indicating a very low affinity for electrons, to 4·0 for fluorine, the most electronegative element, which has a high capacity for attracting electrons. As was the case with atomic radii we find two general trends. The electronegativity increases with atomic number within a given period and, in general, decreases with increasing atomic number within a given group.

3.5. Chemical behaviour — non-transition elements

We learned in Chapter 1 how chemical bonds were formed by the rearrangement of electrons to give the most stable configuration. We can now see how the periodic table helps us to classify the behaviour of elements and compounds whose properties are largely determined by the nature of their chemical bonds.

If we neglect for the moment the transitional elements, we see in the groups of the periodic table a *valency classification*. The group I elements all have a single valency electron and all are strongly electropositive. These elements all have a valency of one and very readily form ionic bonds because of their lack of affinity for electrons. Most of the compounds of these elements are ionic in nature and show characteristic properties. Ionic compounds in general are hard solids with high melting and boiling points. In the solid state they act as insulators but in the molten state or in solution are good conductors of electricity. These properties are a consequence of the nature of the forces acting to hold the structure together in the solid state (see Chapter 6). All of these elements are soft, low melting metals and are chemically extremely reactive.

All the group II elements have two valency electrons and are strongly electropositive. These elements have a valency of two and generally form ionic compounds. Again all the elements are fairly soft metals and are highly reactive.

The elements of group III each have three valency electrons and, with the exception of boron, are only weakly electronegative. These elements generally form trivalent compounds whose bonds have only a small percentage of ionic character. This is so because very large amounts of energy are required to remove three electrons from an atom. Thus covalent compounds are formed which have in general much lower melting and boiling points than ionic materials. With the exception of boron the elements are metals, and boron itself does exhibit some metallic properties.

The chemical behaviour of group IV and V elements is more varied than that in any other group. In this part of the periodic table we find elements of intermediate electronegativity which range from non-metals in the early members of each group to typical metals in the later members. Carbon, silicon and germanium form covalent compounds, exhibiting a valency of four. Tin and lead, on the other hand, whilst forming some tetravalent compounds, also form ionic salts containing divalent ions. Group IV elements are relatively unreactive, although paradoxically carbon probably forms more compounds than any other element. This is because carbon has a facility for forming complicated chain and ring structures to a much greater extent than other elements as is described in greater detail in Chapter 19.

The overall trend in group V is similar to that in group IV, with nitrogen and phosphorus behaving as non-metals, bismuth being typically metallic and arsenic and antimony showing some metallic properties. Apart from bismuth the elements react to form covalent compounds. The group valency can be either three or five, and although nitrogen exhibits only a valency of three the other members of the group form compounds of both types, e.g. PCl_3 and PCl_5.

The group VI elements with the exception of polonium are in general fairly strongly electronegative and are classed as non-metals although tellurium does exhibit some metallic properties. The group valency is two, the bonding being either ionic or covalent depending on the nature of the other element. Thus with weakly electronegative elements typical ionic compounds containing the X^{2-} ion are formed, while in combination with, e.g. hydrogen, the compounds are covalent. Oxygen apart, the group VI elements can exhibit valencies of four and six.

The group VII elements, whose electronic configurations are only one short of the inert gas structure are strongly electronegative and highly reactive. Fluorine, for example, attacks most other elements even at room temperature. The halogens normally exhibit a valency of one, the type of bond depending on the other element present. The halides of electropositive metals are typical ionic compounds whereas their compounds with hydrogen are covalent.

The group VIII elements, commonly described as the *inert gases*, show a great reluctance to take part in chemical reactions. In recent years a small number of inert gas compounds have been successfully prepared, showing that these elements are not quite so unreactive as was once believed.

3.6. Chemical behaviour — transition elements

We have yet to consider the transition elements, i.e. those elements with inner *d*-orbitals incompletely filled, and we shall for the moment include the group zinc, cadmium and mercury under this general heading. The properties of these elements, which are all metals, are quite distinctive and obviously different from those of the typical non-transitional metals. These differences are a direct consequence of the rather unusual electronic configurations of these elements.

The first striking characteristic of the transition metals is the similarity of chemical behaviour horizontally along the period. The reason for this is that although electrons are added as we move along the series, they are accommodated in the incomplete inner *d* orbitals, leaving the important outer electronic configuration unchanged. Since, to a large extent, it is the outer electrons which determine chemical behaviour, we find adjacent elements behaving very similarly.

Another distinguishing feature of these metals is that they have a variable valency. While it is true that some non-transitional elements also have this ability, it is far more pronounced in the transition series. Titanium, for example, forms three different chlorides, $TiCl_2$, $TiCl_3$ and $TiCl_4$. The most common valency encountered is that of two (e.g. $FeCl_2$, $CoCl_2$) since this involves bonds using the two $4s$ outer electrons, but as there is little difference in energy between the $4s$ and $3d$ electrons, these inner electrons readily take part in bond formation as well.

Yet another characteristic of this series of metals is that they have a strong tendency to form *co-ordination compounds* containing complex ions, e.g. $K_4Fe(CN)_6$. Again this is due to the incompletely filled d-orbitals which can be used to accommodate pairs of electrons from other chemical groups and so form chemical bonds. This property can be one of great importance. Although gold is a very unreactive metal and is insoluble in common acids it can be dissolved in *aqua regia* (3 volumes concentrated hydrochloric acid + 1 volume concentrated nitric acid) because of the formation of the complex ion $[AuCl_4]^-$.

Finally, we find that the majority of transition metal salts are highly coloured. Again this is not a property exclusive to transition metal compounds but is fairly characteristic and is due to transitions involving unpaired d-electrons. Compounds of zinc and scandium, whose ions do not contain unpaired d-electrons, are colourless.

CHAPTER 4

The Gaseous State

4.1. The kinetic theory of matter

There are basically three states of matter — the solid, liquid and gaseous states. It is not unreasonable to suppose that the observed properties of a particular state are likely to be related to and indeed a consequence of the way in which the individual units behave, be they molecules, atoms or ions.

To this end, let us imagine the changes that occur at the molecular level in the gradual transition of solid to liquid to gas. In water, for example, the basic unit is the molecule H_2O. X-ray examination reveals that the water molecules in ice are positioned in space, in a highly ordered fashion. A still more detailed analysis reveals that these molecules are not stationary but are vibrating about a mean position. As the temperature is raised the vibrations become more energetic until at 0°C melting of the ice occurs, and liquid water is formed. This sudden change of state that occurs is due to the vibrations becoming so violent that the molecules, although still exerting an attractive influence on each other, are now free to alter their position. We can express this alternatively by saying that molecules in the liquid state have a greater *degree of freedom* and are more disordered than molecules in the solid.

As the temperature is increased beyond 0°C the molecules, in absorbing the energy supplied to them as heat, move about still more rapidly; the system becomes even more disordered, until at 100°C the molecules are now sufficiently energetic to overcome the attractive forces and the liquid boils producing a gas, viz. steam. In this state the molecules exert only a very weak mutual attraction and as a result are free to move about anywhere within the container; the continual collisions occurring between the molecules and the walls of the container create the effect we describe as pressure. Clearly, the gaseous state is much less ordered than the liquid state with the molecules having a much greater degree of freedom.

We must now consider whether the theory is consistent with experimental observation; like any theory it must be put to the test.

TEST I. *The absolute zero of temperature.* In describing what we mean by a kinetic theory of matter we have identified heat with molecular motion for we have presumed that the effect of increasing the temperature increases this motion and produces a consequent increase in disorder; i.e. we can say that not only is temperature a measure of the 'heat content' of a system, but is indeed a measure of the average kinetic energy of the molecules. We might well feel tempted at this point to pose the question 'to what value may the temperature be reduced?'

It follows from our previous remarks that if heat is removed from a system its temperature will fall, molecular motion will become less vigorous and a more ordered state will result. Inevitably a temperature must be reached when the system assumes a state of maximum order.† Thus, on the basis of the kinetic theory, *an absolute zero of temperature* is predicted; Lord Kelvin deduced that for *any* substance, this temperature is $-273 \cdot 15°C$. Although this temperature has never been reached, experiments have been performed where, for a short period of time, the temperature has been lowered to $-273 \cdot 1499°C$.

---

†At first sight we might reasonably expect a condition of maximum order to be associated with a complete cessation of molecular motion. However, modern physics requires that molecular motion can never cease entirely.

It is convenient to introduce an *absolute temperature scale* which has as its zero-point the absolute zero of temperature. Any temperature on this scale is indicated by the symbol T and is expressed in °K (Kelvin). Thus, if the size of the absolute degree is made the same as the centigrade degree it follows that

$$T(°K) = t(°C) + 273 \text{ (approx.)}$$

TEST II. *Brownian Movement.* The behaviour of very small pollen grains suspended in water provides a striking confirmation of the kinetic theory. These tiny particles, when examined under a microscope, are observed to be in constant motion, moving about in a completely unpredictable fashion. This chaotic motion, referred to as the *Brownian Movement* (after the botanist Robert Brown who first observed the phenomenon), can only be attributed to the fact that the molecules of the liquid in which the pollen particles are suspended are moving about in a similarly chaotic, random manner. Thus, at one instant of time, a small pollen particle may be struck by, say, three water molecules in a manner as shown in Fig. 4.1(a). The net result of this molecular bombardment would be to

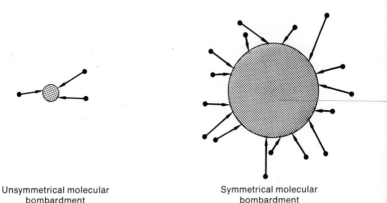

Unsymmetrical molecular
bombardment
(a)

Symmetrical molecular
bombardment
(b)

FIG. 4.1. Brownian movement.

cause it to move off in a new direction. Over a period of time, the particle would experience many changes of direction.

A similar effect can be seen if smoke is examined under a microscope. In this case the denser solid particles experience the Brownian movement and for this reason do not settle out under the influence of gravity.

As the particle size increases the particle motion becomes less violent and indeed for large particles the Brownian movement is completely absent. That this is so is due to the fact that the large particles are more likely to suffer a symmetrical molecular bombardment in which molecular collisions neutralize each other (Fig. 4.1(b)).

4.2. Gas Laws

Charles' Law.

The pressure exerted by a gas when measured under the condition of constant volume is found to vary linearly with its temperature (°C). This relationship, expressed graphically in Fig. 4.2, implies that if the gas continued to behave in the manner indicated by the unbroken line, then its pressure would become zero at a temperature of −273°C. Now in the light of the kinetic theory, pressure is the result of molecular collisions

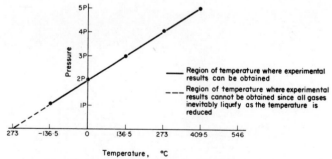

FIG. 4.2. Variation of gas pressure with temperature.

with the walls of the container; zero pressure must therefore mean a total absence of molecular motion and the temperature corresponding to this state must be an absolute zero of temperature. Thus, *for a fixed amount of gas under the condition of constant volume the pressure is directly proportional to the absolute temperature;* this statement is one form of Charles' law,

i.e. $P \propto T$ (fixed amount of gas at constant volume)

These observations by Charles are entirely consistent with the kinetic theory, for as the temperature is raised, molecular velocity will increase and molecular collisions with the container walls will be both more frequent and more vigorous, giving rise to a greater pressure.

Boyle's law

The relationship between gas pressure and volume was investigated by Robert Boyle, who was able to show that, *at constant temperature, the pressure exerted by a given amount of gas varies inversely with its volume* (Fig. 4.3),

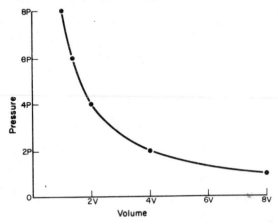

FIG. 4.3. Variation of gas pressure with volume.

i.e. $P \propto 1/V$ (fixed amount of gas at constant temperature). This relationship, referred to as Boyle's law, is exactly what we would expect on the basis of our theory. If the temperature remains constant the average velocity of the molecules likewise does not change and if the volume of the container is halved, the number of molecules striking a unit area of surface per second must double, i.e. the pressure must double.

Graham's law

Gases have the ability to diffuse through porous materials. Graham in 1883 studied this aspect of gaseous behaviour and showed that *the rate of diffusion was inversely proportional to the square root of the molecular weight of the gas* (Graham's law),

i.e. $$\text{rate of diffusion} \propto \sqrt{\frac{1}{M}}$$

where M = molecular weight.

On the basis of the kinetic theory, it seems reasonable to suppose that the rate of diffusion will be determined by the average speed of the molecules. Consider therefore two gases A and B, each at the same temperature and therefore each being associated with the same amount of kinetic energy per molecule. Now if c_A and c_B are the respective average velocities, with m_A and m_B, the respective molecular masses, then it follows that for the two gases at the same temperature

$$\tfrac{1}{2}m_A C_A^2 = \tfrac{1}{2}m_B C_B^2 \quad (\text{kinetic energy} = \tfrac{1}{2}mC^2)$$

$$\therefore \qquad \frac{C_A}{C_B} = \sqrt{\frac{m_B}{m_A}}$$

The molecular weight of a gas is the weight of a particular number of molecules. This number, referred to as the *Avogadro Number* (N), is very large and has a value of $6 \cdot 023 \times 10^{23}$.

$\therefore \qquad M_A$ = molecular weight of gas A = Nm_A

and $\qquad M_B$ = molecular weight of gas B = Nm_B

$$\therefore \qquad \frac{C_A}{C_B} = \sqrt{\frac{M_B}{M_A}}$$

i.e.   Average molecular velocity $\propto \sqrt{\frac{1}{M}}$

4.3.   Ideal gas equation of state

The laws of Boyle and Charles are found to be only truly valid under conditions of low pressure and high temperature. Under these conditions the gas is said to behave ideally and is described as an *ideal gas*. Under conditions for which the laws are not accurately obeyed the gas is said to be non-ideal. For the moment, we will restrict ourselves to the consideration of an ideal gas.

The combination of Boyle's law,

$$P \propto \frac{1}{V} \text{ (fixed amount, constant temperature)}$$

and Charles' law,

$$P \propto T \text{ (fixed amount, constant volume)}$$

yields

$$P \propto \frac{T}{V} \text{ (fixed amount)}$$

Hence $PV/T = k$ where k, a constant of proportionality, will have a value appropriate to the amount and nature of the gas being considered.† If this amount is chosen to be one gram then k will have a value specific to a particular gas (Table 4.1).

Thus for *any* weight (w) of gas

$$\frac{PV}{T} = wk$$

This relationship, termed an *equation of state*, relates the amount of gas w to its pressure, volume and temperature. The above equation, though useful, suffers from the disadvantage of having

†The numerical value of k will also depend on the units of pressure and volume employed; this point will be discussed later.

TABLE 4.1

Gas	$\dfrac{PV}{T} = k$ (1. atm deg^{-1} g^{-1})	kM (1. atm deg^{-1} mole^{-1})
Hydrogen	0·041	0·082
Oxygen	0·00256	0·082
Nitrogen	0·00293	0·082

different k values for different gases. This restriction can be eliminated in the following manner:

$$\frac{PV}{T} = wk$$

Multiplying top and bottom by M, the molecular weight, gives

$$\frac{PV}{T} = \frac{w}{M} \, (kM)$$

Now the number of moles of gas, $n = w/M$

$$\therefore \qquad \frac{PV}{T} = n \, (kM)$$

Both k and M are constants characteristic of a particular gas. Their product, however, is the same for all gases (Table 4.1); it is called the *gas constant* and is given the symbol R, where $R = kM$.

$$\text{Thus } PV = nRT$$

This equation is the *ideal gas equation*.

The fact that the product of k and M is a constant follows as a consequence of *Avogadro's law* which states that *equal volumes of all gases measured under the same conditions of temperature and pressure, contain an equal number of molecules.* That this is so can be shown in the following way:

Consider two gases A and B. Then

for gas A, $\quad \dfrac{P_A V_A}{n_A T_A} = R$ and for gas B, $\dfrac{P_B V_B}{n_B T_B} = R$

If $P_A = P_B$ and $T_A = T_B$

then $\dfrac{V_A}{n_A} = \dfrac{V_B}{n_B}$

where $n_A = \dfrac{\text{no. of A molecules}}{N \text{ (Avogadro number)}}$

and $n_B = \dfrac{\text{no. of B molecules}}{N}$

$\therefore$ $\dfrac{V_A}{\text{no. of A molecules}} = \dfrac{V_B}{\text{no. of B molecules}}$

Now if V_A and V_B are made equal then the above equality is valid only if the number of A molecules equals the number of B molecules.

The numerical value of R may be determined by inserting into the ideal-gas equation known values of n, P, V and T. Thus, under standard conditions of temperature and pressure (S.T.P. conditions), i.e. at 0°C and 1 atm, one mole of *any* gas occupies a volume of 22·414 litres.

$\therefore \qquad R = \dfrac{1 \, (\text{atm}) \times 22 \cdot 4 \, (1)}{273 \, (\degree K) \times 1 \, (\text{mole})}$

$= 0 \cdot 082$ litre atmosphere per degree per mole

$= 0 \cdot 082$ 1 atm deg^{-1} mole^{-1}

This, of course, is not the only value R may assume, for the above choice of pressure and volume units is arbitrary. For example, if the pressure is expressed in units of dynes cm^{-2} and volume as cm^3

then $R = \dfrac{1 \cdot 014 \times 10^6 (\text{dynes cm}^{-2}) \times 22{,}400 \, (\text{cm}^3)}{273 \, (\degree K) \times 1 \, (\text{mole})}$

$= 8 \cdot 312 \times 10^7$ dynes cm deg^{-1} mole^{-1}

$= 8 \cdot 312 \times 10^7$ ergs deg^{-1} mole^{-1} (1 dyne cm $=$ 1 erg)

The units of the gas constant are most clearly demonstrated in this way to be those of work. The most commonly used unit of energy in chemistry is the calorie and since 1 cal = 4·18 × 10^7 ergs, it follows that $R = 1·987$ cal deg^{-1} mole^{-1}.

The ideal-gas equation is one whose use in physical chemistry is considerable. The following example illustrates some of the simple manipulations which must be performed in the use of the equation.

EXAMPLE 4.1. *Calculate the number of molecules remaining in an electronic vacuum tube of capacity 180 cm³ which has been evacuated and sealed off at 25°C to a pressure of 1·0 × 10⁻⁷ mm Hg.*

$$PV = nRT$$

$$\therefore \quad \frac{10^{-7}}{760}(\text{atm}) \times \frac{180}{1000}(1) = n \times 0.082 \times 298$$

$$\therefore \quad n = 9.694 \times 10^{-13} \text{ mole}$$

Number of molecules = number of moles × Avogadro No.

$$= 9.694 \times 10^{-13} \times 6.023 \times 10^{23}$$
$$= 5.838 \times 10^{11}$$

Note that considerable care must be exercised in the choice of units employed for pressure and volume. If the value of the gas constant chosen is 0·082 1 atm deg^{-1} mole^{-1}, then pressure and volume *must* be expressed in units of atmospheres and litres respectively.

4.4. Gas mixtures

Not infrequently the chemist is faced with a problem involving not one single gas but a gas mixture. In order to tackle such a problem it is necessary to know what contribution will be made to the total pressure by each gas in the mixture, relative to the amount of that particular constituent present. This contribution is termed the partial pressure of the gas and is defined as being *the pressure it would exert if it alone occupied the entire volume of the container at the specified temperature.*

Consider now a mixture of gases A and B for which the measured (i.e. total) pressure is P. If now we imagine gas B to be removed the pressure within the container will be that due to gas A, i.e. the partial pressure of A.

From the ideal-gas equation

$$P_A = \frac{n_A RT}{V}$$

where P_A = partial pressure of A and n_A = no. of moles of A. Similarly, the pressure within the container upon the imaginary removal of gas A is the partial pressure of B,

i.e.
$$P_B = \frac{n_B RT}{V}$$

where P_B = partial pressure of B, n_B = no. of moles of B.

For the gas mixture:

$$P = \frac{n_T RT}{V}$$

where n_T = total moles present = $n_A + n_B$.

$\therefore$
$$P_A = \frac{n_A \cdot P}{n_T} \quad \text{and} \quad P_B = \frac{n_B \cdot P}{n_T}$$

The fractions n_A/n_T and n_B/n_T are described as being the *mole fractions* of A and B respectively. The symbol x is the one most commonly used to represent a mole fraction. Thus

$$\frac{n_A}{n_T} = x_A \quad \text{and} \quad \frac{n_B}{n_T} = x_B$$

$\therefore$
$$P_A = x_A \cdot P \quad \text{and} \quad P_B = x_B \cdot P$$

It is now possible to see that the contribution made by each gas in a mixture to the total pressure is determined by the relative amount of the gas present. Thus, if the gases A and B are present in the mole ratio of $1:3$ respectively then

$$P_A = \frac{P}{4} \quad \text{and} \quad P_B = \frac{3P}{4}$$

The relationship between the measured total pressure and the respective partial pressure can also be rapidly deduced, for

$$P_A = \frac{n_A \cdot P}{n_T} \quad \text{and} \quad P_B = \frac{n_B \cdot P}{n_T}$$

$$\therefore \quad P_A + P_B = \left(\frac{n_A}{n_T} + \frac{n_B}{n_T}\right) P$$

$$= \left(\frac{n_A + n_B}{n_T}\right) P$$

$$= P$$

Thus, *the total pressure of a mixture of gases is the sum of the individual partial pressures of the gases in the mixture.* This statement is *Dalton's law of partial pressures.*

EXAMPLE 4.2. *A volume 0·54 l. of argon (at wt. 40) measured at 72 cm Hg and 15°C together with 2·32 g of neon (at wt. 20) are placed in a 2-litre flask. What will be the partial pressures of the two gases when the gas mixture is brought to a temperature of 100°C?*

Method I

$$\text{No. of moles of argon} = \frac{PV}{RT} = \frac{72 \times 0·54}{76 \times 0·082 \times 288}$$

$$= 0·02166$$

$$\text{No. of moles of neon} = \frac{\text{wt. of neon}}{\text{at. wt. of neon}} = \frac{2·32}{20}$$

$$= 0·116$$

$$\therefore \quad \text{Total no. of moles} = 0·1377$$

$$PV = nRT$$

$$\therefore \quad P = \frac{0·1377 \times 0·082 \times 373}{2}$$

$$= 2·106 \text{ atm}$$

$$\therefore \qquad P_{\text{argon}} = x_{\text{argon}} \cdot P = \frac{0 \cdot 02166}{0 \cdot 1377} \times 2 \cdot 106 = 0 \cdot 3313 \text{ atm}$$

$$\therefore \qquad P_{\text{neon}} = x_{\text{neon}} \cdot P = \frac{0 \cdot 116}{0 \cdot 1377} \times 2 \cdot 106 = 1 \cdot 774 \text{ atm}$$

Method II

In this alternative method, use is made of the relationships

$$P_{\text{argon}} = \frac{n_{\text{argon}} RT}{V} \qquad \text{and} \quad P_{\text{neon}} = \frac{n_{\text{neon}} RT}{V}$$

$$= \frac{0 \cdot 02166 \times 0 \cdot 082 \times 373}{2} \qquad = \frac{0 \cdot 116 \times 0 \cdot 082 \times 373}{2}$$

$$= 0 \cdot 3313 \text{ atm} \qquad \qquad = 1 \cdot 774 \text{ atm}$$

4.5. Energy and speed distributions

The molecules in a sample of gas do not all possess the same kinetic energy nor therefore the same speed. (Kinetic energy per molecule $= \frac{1}{2} mc^2$.) There is in reality a *distribution of both energy and speed*. This distribution arises as a direct consequence of the chaotic motion of the molecules. Let us suppose that it was possible to introduce into a container gas molecules all of which initially were travelling at the same speed and therefore possessing the same kinetic energy. This situation would soon alter, for some molecules would, during collision, lose kinetic energy thereby reducing their speed while others would experience a gain in kinetic energy with a consequent increase in speed. Although during a collision between two molecules, there is likely to be an exchange of energy (one molecule gaining in kinetic energy at the expense of the other) the total kinetic energy is conserved. Such a collision is described as being perfectly elastic. If such were not the case then, as a result of continual collision, all the molecules would experience a gradual decrease in speed and would inevitably settle out to the foot of the container; that this does not happen is due to the elastic character of the collisions.

The manner in which the molecules distribute themselves over all possible speeds is shown in Fig. 4.4(a). A somewhat different shape of curve is given for the corresponding energy distribution (Fig. 4.4(b)).

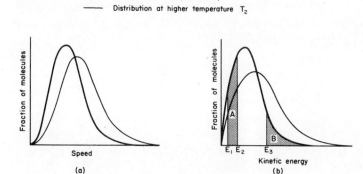

FIG. 4.4. Speed and energy distributions.

Examination of Fig. 4.4(a) reveals that

(i) Very few molecules have either extremely low or high speeds, and

(ii) there is one speed which is more probable than any other, this speed corresponding to the maximum in the curve.

Similar deductions apply to Fig. 4.4(b).

Any increase in temperature will increase both the average kinetic energy and the speed of the molecules and although the general shape of the above curves remains the same, there is a displacement of the maxima towards higher speeds and higher energies.

One final point of interest with regard to the above diagrams is to consider the significance of various areas under the curve. In Fig. 4.4(b), the area A, relative to the total area under the curve, represents the fraction of molecules having energy values between E_1 and E_2. Of greater interest is the fraction of molecules which have energies in excess of a particular value. Thus, area B is a measure of the fraction of molecules having energy values between E_3 and infinity, i.e. having energies in excess of E_3.

When we come to consider evaporation (Chapter 5) and reaction rates (Chapter 12) we will learn that it is these highly energetic molecules which to a large extent control these processes.

4.6. Real gases

An ideal gas is one in which (a) the molecules are regarded as point masses, and (b) attraction between the molecules is ignored. In reality the molecules of a gas *do* experience a mutual attraction and *do* occupy space, small though it may be. As a result, the ideal generalization $PV = nRT$ is hardly valid when the behaviour of gases is to be studied accurately. Deviations from ideal behaviour are conveniently stated in

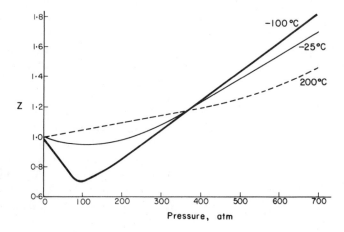

Fig. 4.5. Variation of Z with pressure and temperature.

terms of the *compressibility factor*, which is defined by

$$Z = PV/nRT$$

For an ideal gas, $Z = 1$, and any deviation from ideal behaviour will be measured by the departure of the value of Z from unity. In Fig. 4.5, Z is plotted against the pressure for the gas nitrogen at various temperatures. Clearly, the extent of the deviations from ideality is dependent upon both the pressure and temperature. Before any attempt is made to account for this pattern of behaviour some knowledge of the nature of attractive forces is required.

4.7. Molecular interactions

Interactions occurring between molecules are electrostatic in origin and may be either repulsive or attractive. The repulsive interaction principally arises through the overlap of electron clouds and only becomes appreciable at very close intermolecular separations; attractive interactions are similarly dependent on the distance of separation and only when the latter is extremely small does the repulsive effect exceed its attractive counterpart.

Attractive interactions are set up between all molecules, polar or non-polar, and are referred to as *van der Waals* interactions. The origin of this interaction for polar molecules is easy to appreciate for within such a molecule the centres of positive and negative charge do not coincide, and this condition leads to a situation in which the preferred configuration is one of attraction, with the positive end of one molecule tending to align itself with the negative end of a second and so on.

The origin of attractive forces between non-polar molecules is less clear. While such molecules possess no permanent dipole moment, there will very probably be at any instant of time a separation of the centres of positive and negative charge. This *instantaneous* dipole, arising through the continual movement of the electrons, can induce an electric dipole in a nearby molecule and thereby create an attractive interaction. At some later moment in time there will be a new electron distribution but again an overall attraction results through the interaction of the instantaneous dipole of one molecule with the induced dipole of a neighbouring molecule.

4.8. The van der Waals treatment of a real gas

The first attempt to improve the simplified model of the ideal gas was made by the Dutch scientist van der Waals in 1873. In the ideal gas, V is identified with the volume of the container; in reality, it represents the *space* in which the molecules are free to move, and account must be taken of the fact that some of this space is unavailable for this purpose, is in fact 'excluded', due to the finite volume of the molecules. Hence, the space in which the molecules are free to move corresponds to a volume somewhat less than that of the container, and V in the ideal gas equation must be replaced by $V-b$, where V is the molar volume and b, the excluded volume, corresponds to the effective volume of all the molecules in one mole. Thus, the first modification of the ideal gas model results in the equation,

$$P(V-b) = RT \text{ for one mole of gas}$$

An examination of this equation, reveals the nature of the deviation produced by the finite volume of the molecules, for

$$\frac{PV}{RT} = Z = 1 + \frac{Pb}{RT}$$

Hence, the deviation produced as a result of this volume effect causes the value of Z to increase its value in excess of unity, and the greater the pressure, the greater will be the value of Z.

The second correction made to the ideal gas equation results from a consideration of the intermolecular attractive forces. The way in which these forces produce deviation from ideal behaviour is illustrated in the following diagram (Fig. 4.6). A molecule within the bulk of the gas experiences no resultant

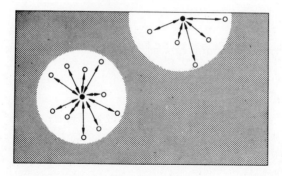

Fig. 4.6. Molecular attractions in the bulk of the gas and at the container wall.

attractive force because the field of force surrounding this molecule is spherically symmetrical. This symmetry is absent when the molecule is about to strike the wall of the container, and such a molecule will consequently be under the influence of a net inward pull. This effect will decrease the pressure below that expected for an ideal gas and it therefore becomes necessary to add to the observed pressure a correction term in order to obtain that pressure an ideal gas would exert under the same conditions. It can be shown that the force of attraction is proportional to the square of the density and thus inversely proportional to the square of the volume,

$$\text{i.e. correction term} \propto \frac{1}{V^2} = \frac{a}{V^2}$$

where V is again the molar volume and a is the constant of proportionality whose magnitude gives a measure of the strength of the intermolecular forces. Neglecting for the moment the correction previously made to V in the ideal gas equation, the effect of molecular attraction would alter the latter to

$$\left(P + \frac{a}{V^2}\right)V = RT \text{ for one mole of gas}$$

Hence $$\frac{PV}{RT} = Z = 1 - \frac{a}{RTV}$$

Thus, the deviation produced in consequence of this effect causes the value of Z to decrease below unity, and as the pressure is increased, the volume will decrease and the value of Z will fall still more below that of unity. This is readily understandable because as the pressure is increased the molecules are separated from each other by smaller distances and the attractive forces increase in strength—as a result, the net inward pull suffered by molecules about to strike the surface is increased.

It can now be appreciated that the two effects producing deviation from ideal behaviour act in opposition to each other with one tending to increase, the other to decrease the value of Z. Whichever effect predominates will depend on the gas, the temperature and the pressure.

The equation which corrects for both the size of the molecules and their mutual attraction is called the *van der Waals* equation of state, viz.

$$\left(P + \frac{a}{V^2}\right)(V - b) = RT \text{ for one mole of gas}$$

We can now explain in semi-quantitative terms the various features of the graphs in Fig. 4.5. Consider the behaviour of nitrogen at $-100°C$. Within the region of pressure in which $Z = 1$, the effect of molecular attraction predominates, whilst at higher pressures, where $Z = 1$, the influence of the excluded volume is the more marked. At exceedingly low pressures, both the correction terms a/V^2 and b are very small compared with P and V, in which case the ideal gas equation is obeyed.

The reason that the deviations caused by van der Waals forces depend on temperature and indeed (considering this effect alone) are more pronounced at low rather than high temperatures can be explained as follows. It will be recalled that heat is identified with molecular motion—as a result of this, the attractive forces are more effectively opposed by fast-moving molecules than by molecules travelling on the average at lower velocities.

Also, it is an observed fact that heavy molecules deviate more than light molecules under the same conditions of temperature and pressure; this is so because all molecules have, at the same temperature, approximately the same kinetic energy. Consequently, heavy molecules will have lower velocities than light molecules and the forces of attraction will have a greater influence on them.

4.9. The critical state of a gas

The forces of attraction existing between molecules, which in the gaseous state give rise to deviations from ideal behaviour, are responsible for the existence of a liquid state, which indeed can be thought of as being an extremely imperfect gas. To bring about the conversion of gas to liquid we must select conditions which will increase the magnitude of the attractive forces. In this context the work of Andrews, carried out in 1869, is of considerable interest. He established the pressure–volume–temperature relationships of carbon dioxide, as a result of which the conditions essential for the liquefaction process were revealed. He showed that unless the temperature of the carbon dioxide was below a certain critical value, sensibly called the *critical temperature* (T_c), liquefaction of the gas was impossible irrespective of what pressure was applied.

Further investigation has shown that this phenomenon is quite general, the limiting temperature above which liquefaction cannot occur depending on the gas being considered. Below this temperature,† the kinetic energy of the molecules is sufficient to allow the forces of attraction to produce liquid when an appropriate pressure is applied. Clearly gases having high 'a' values will also have high T_c values, and in view of the

---

†A gaseous substance below the critical temperature is described as a vapour since it can be condensed to a liquid by pressure alone.

fact that the normal boiling point T_b is approximately related to the critical temperature by the equation

$$T_b \approx \tfrac{2}{3} T_c$$

it follows that the normal boiling point is a measure of attractive forces acting between the molecules.

4.10. Liquefaction of gases

By the early part of the nineteenth century, a large number of gases had been successfully liquefied, but attempts to liquefy hydrogen, oxygen, nitrogen and carbon monoxide (all of which are characterized by low *a* values) failed. The concensus of opinion at that time was that these gases could not be liquefied and were consequently given the name *permanent gases*. The work of Andrews strongly suggested that the failure to liquefy the so-called permanent gases was due to the fact that the temperatures reached were insufficiently low. His contribution acted as a spur to other workers to develop new techniques for obtaining low temperature.

At present, two general principles (or a combination of the two) are employed, to liquefy gases on an industrial scale. The first of these makes use of the *Joule–Thomson effect*. Joule and Thomson in 1854 showed that when a gas under pressure is allowed to pass through a valve or throttle into a region of lower pressure, a cooling effect is observed. This change in temperature, known as the Joule–Thomson effect, is a result of the gas not being ideal, and arises from the energy required to overcome the forces of attraction between the molecules in moving them further apart. If the apparatus is effectively insulated from its surroundings,† the only available source of energy is that due to the thermal motion of the molecules. Thus the kinetic energy of the molecules will decrease and this must result in a fall in temperature. At ordinary temperatures this cooling effect

†An expansion in which heat is neither allowed to enter or leave the system is called an *adiabatic expansion*.

is, for most gases, extremely small, and as the temperature is raised it becomes even smaller until at one particular temperature characteristic of the gas there is no change in temperature at all. The gas is then said to be at its *inversion temperature*. Above this the gas, on expansion through a valve, experiences a heating effect. This rise in temperature is related to the shapes of the high temperature isotherms in Fig. 4.5, where Z continually increases with an increase in pressure due to the effect of the excluded volume predominating over that of molecular attraction.

Hydrogen, helium and neon are the only gases not below their inversion points at ordinary temperatures. Hence, precooling before expansion would be necessary in order to liquefy those three gases by the Joule–Thomson effect.

Linde Process. Carl von Linde in 1895 successfully developed a technique for the liquefaction of air. In this process (Fig. 4.7) air, previously freed of carbon dioxide by passage over caustic soda, is compressed to about 200 atm, and passed through cooling coils in order that the heat generated during the compression be removed. The gas stream then passes through the heat exchanger from which it is expanded to about 1 atm through the expansion valve where it is cooled by the Joule–Thomson effect. The gas is then returned to the compressor via the heat exchanger thus causing the temperature of the incoming compressed gas to be lowered prior to its expansion. The repetition of this cycle results in the temperature of the expanding gas dropping sufficiently to cause partial liquefaction.

The second of the two principles involves cooling by the performance of external work. Consider a gas under pressure enclosed in a cylinder fitted with a piston. When the pressure exerted on the piston is reduced to a value below that of the gas, the latter will expand and cause the piston to move up the cylinder. This performance of work by the gas requires the expenditure of energy. If the expansion is carried out under adiabatic conditions, the performance of mechanical work can

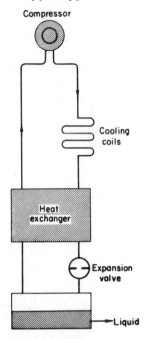

FIG. 4.7. Linde process for liquefaction of air.

only be achieved at the expense of the kinetic energy of the molecules hence the temperature of the gas drops.

Claude Process. The Claude process for the liquefaction of air employs a combination of the two principles discussed. Air is compressed to 200 atm, the heat generated being removed by cold water. It then passes to an expansion engine where, in performing work, it experiences a very considerable drop in temperature for the reason just outlined. If this cycle were repeated several times, liquefaction of the air would result. However, such a process in itself would be highly inefficient because of the formation of liquid in the expansion engine. In practice, therefore, this method of cooling is used only to produce a sufficiently low temperature for liquefaction to occur by the Joule–Thomson effect. Hence, the air from the compressor is divided into two streams (Fig. 4.8). One goes to the

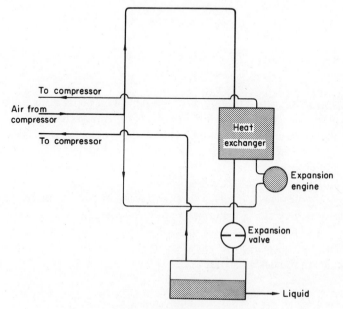

FIG. 4.8. Claude process for liquefaction of air.

expansion engine where it is cooled. It is returned to the compressor via the heat exchanger where it precools the second stream of air, the temperature of which is further reduced by its passage through the expansion valve at which point partial liquefaction results. The Claude process is economically more attractive than the Linde process, for the energy released by the gas during its expansion is recoverable. No such recovery is possible in the Linde process; it has, however, the attraction of having no moving parts — this is very much an advantage when we bear in mind that the possibility of mechanical failure is greater at lower temperatures.

CHAPTER 5

The Liquid State

A LIQUID can be conveniently thought of as being a continuation of the gas phase into a region of smaller volume and higher molecular attraction. The attractive forces are sufficiently powerful to prevent the liquid from filling the whole of the containing vessel (as would a gas) but too weak to allow the liquid to have any characteristic shape (as in the case of a solid).

Although the attractive forces in a liquid restrict molecular motion, they do not eliminate it. The molecules of the liquid are in continuous random motion, their velocities and kinetic energies following the same distribution pattern as for molecules in the gaseous state.

5.1. Surface tension

In Chapter 4 we saw that a molecule in the gaseous state, about to strike the wall of the container, is under the influence of a net inward pull. Because of this, the observed pressure is less than that which an ideal gas would exert under the same conditions.

A similar situation arises at the liquid surface with the surface molecules being attracted inwards as a result of the imbalance in the intermolecular attractions at the surface (Fig. 5.1). Because of this purely surface effect, the surface molecules behave differently from those in the bulk of the liquid. The difference manifests itself in a number of interesting ways. For example, a drop of liquid tends to assume a shape which presents the smallest surface area for a given volume

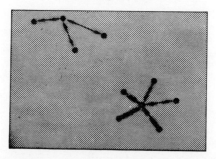

FIG. 5.1. Molecular attractions in the bulk of a liquid and at
the surface.

of liquid, viz. a sphere. In this way the number of surface
molecules is minimized. Also, certain objects such as a razor
blade or a needle can be made to float on the surface of water
even though they are very much more dense than water. These
phenomena demonstrate that the surface behaves very much
like an elastic membrane under tension, tending in the case of
a liquid drop to reduce its area to a minimum and, in the case
of the razor blade, to resist penetration of the surface. For this
reason the term *surface tension* is used to describe the effect.

Because surface tension is a consequence of intermolecular
attractions, it is not surprising that surface tension decreases
with increase in temperature, for as the temperature rises,
forces of attraction grow weaker. Surface tension can also
be reduced by adding "wetting agents". Thus the solution of
a soap in water lowers the surface tension of water thereby
increasing its cleansing action (Chapter 28).

5.2. Viscosity

Viscosity is the resistance to flow of a liquid. Thus water
shows a greater tendency to flow than syrup and we describe
water as being less viscous than syrup.

What explanation can be offered for this particular aspect of
liquid behaviour? The flow process is one which involves

molecules sliding past each other under the influence of some applied stress. The rate of flow will depend upon (i) the magnitude of the stress, (ii) the shape of the molecules, and (iii) the magnitude of the forces of intermolecular attraction. Thus, molecules approximately spherical in shape will more readily slide past each other than, e.g. long chain polymer molecules which can become entangled in each other. Also, molecules among which attractive forces are weak will flow more easily than molecules strongly bound to each other.

The viscosity of a liquid nearly always decreases with increase in temperature. This again can be thought of as a consequence of the way in which temperature influences forces of molecular attraction.

5.3. Evaporation

No matter how strong may be the attractive forces in a body of liquid, there will always be some molecules which will overcome these forces, leave the liquid and continue their existence in the gaseous state. Such a transformation of liquid into vapour is referred to as evaporation. The amount of energy in the form of heat that is required to bring about the evaporation of a stated amount of liquid (either 1 gram or 1 mole) is the *latent heat of vaporization*, the magnitude of this quantity being yet another measure of the strength of molecular attraction.

The mechanism by which liquids evaporate is explained on the basis of the kinetic theory. We have already remarked that molecules in the liquid state have their energies distributed over a wide range of values as in Fig. 5.2. At any one instant of time some surface molecules will, through collision, be directed upwards and out of the liquid. The future of such a molecule then depends on how energetic it is at the moment of its departure, for only molecules which possess kinetic energy in excess of a particular value will be sufficiently energetic to overcome the attractive forces operative in the liquid state and

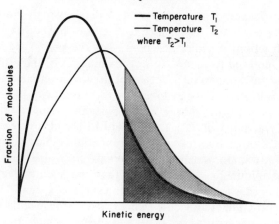

Fig. 5.2. Influence of temperature on the distribution of molecular energies.

hence leave the surface. Thus in Fig. 5.2, if E represents this critical value of kinetic energy, the shaded area is a measure of the fraction of molecules which possess kinetic energy in excess of this, and the unshaded area corresponds to the fraction of molecules which does not meet the energy requirements for evaporation. In short, two conditions must be satisfied if a molecule is to evaporate:

(i) It must occupy a position at the surface.

(ii) It must have kinetic energy in excess of a particular value.

5.4. Rate of evaporation

Extending the above argument, it follows that the rate of evaporation, expressed by the number of molecules leaving the liquid surface per minute, will be dependent upon two factors:

(i) *The amount of surface presented by the liquid*—the greater the surface area, the greater is the number of molecules occupying positions at the surface.

(ii) *The fraction of molecules having kinetic energy in excess of the critical amount.* This quantity will be strongly influenced by the temperature. Reference to Fig. 5.2 shows that a much larger fraction of molecules possesses the necessary kinetic energy required to escape from the influence of the attractive forces at the temperature T_2 than at the lower temperature T_1. As a result, the rate of evaporation increases with increase in temperature. Furthermore, for one particular temperature, this critical fraction will vary from liquid to liquid. For example, molecules of carbon tetrachloride experience only very weak mutual attraction (small van der Waals "a" value) and therefore the minimum kinetic energy that one such molecule requires in order to break free from the surface is small. As a result, the fraction of molecules which, at any one instant of time, meets the necessary energy requirements is high and so the rate of evaporation will likewise be high.

5.5.   Evaporation and cooling

Consider a liquid evaporating in an open vessel, so that vapour molecules can diffuse into the atmosphere. If only the more energetic molecules are those lost by the liquid then it follows that the average kinetic energy of the remaining molecules must decrease. As heat is the kinetic energy of molecular motion the temperature of the liquid must therefore fall. *There is thus an attendant cooling effect accompanying evaporation*, the magnitude of which will depend on how effectively insulated is the liquid from its surroundings. If, for the moment, we imagine that no heat enters the system from the surroundings, then cooling would continue until a temperature was reached at which virtually no molecules would possess sufficient kinetic energy to break free of the surface. In this condition the evaporation rate would be zero. In practice this situation can never be realized completely, as there must always be at least a slight heat flow from the surroundings no matter how well insulated the vessel may be.

5.6. Evaporation in a closed container

Attention must now be drawn to the possibility of vapour molecules returning to the liquid state — the *condensation process*. The extent to which this process can superimpose itself on evaporation depends on whether the liquid is evaporating in an open or a closed vessel. In an open vessel, where vapour molecules are free to diffuse away from the liquid surface, condensation does not occur to an appreciable extent. However, in a closed vessel, where vapour molecules are not in a position to leave the system completely, the condensation process can no longer be ignored.

Consider therefore a vessel from which all air has been removed and into which there is injected liquid. Initially the pressure within the container will be zero. However, as a result of evaporation the pressure will increase as more and more molecules enter the space above the liquid. At any one instant of time, some of these molecules will through sheer chance collisions be directed back to the liquid surface. Thus the two opposing processes of evaporation and condensation proceed simultaneously.

If a condition of constant temperature is imposed on the system, the fraction of liquid molecules possessing kinetic energy in excess of the critical value will be constant and evaporation will therefore proceed at a constant rate (Fig. 5.3). On the other hand, the rate of condensation will be proportional to the number of molecules occupying the space above the liquid. This number, initially zero, increases up to a limiting value at which point the number of molecules leaving the surface equals the number returning at any one instant of time,

i.e.   Rate of evaporation = Rate of condensation

Such a system is described as being in a condition of *dynamic equilibrium*, the term dynamic being used to emphasize the fact that the equilibrium is by no means static. In this equilibrium state, the pressure exerted by the vapour is referred to

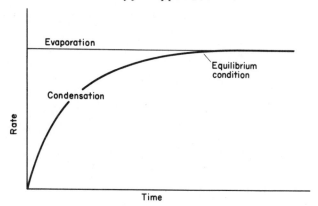

Fig. 5.3. Variation of evaporation and condensation rates with time.

as the *vapour pressure of the liquid* and is defined as being *the pressure the vapour exerts when liquid and vapour co-exist in a condition of dynamic equilibrium.* Its value is dependent on the nature of the liquid and upon the temperature.

5.7. Variation of vapour pressure with temperature: boiling

The fraction of molecules capable of leaving a liquid surface increases rapidly with temperature. The result, true for all liquids, is a rapid increase in the vapour pressure with increase in temperature.

When the vapour pressure of a liquid equals the external pressure, the liquid is said to boil. Thus, water boils at 100°C under a pressure of 1 atm because at this temperature the vapour pressure of water is 1 atm. If the external pressure is increased the water will no longer boil at 100°C but at some temperature in excess of this value. Similarly, if the external pressure is reduced below 1 atm, water can be made to boil at temperatures lower than 100°C.†

†Because the boiling point of a liquid varies with the external pressure, it is necessary, when quoting a boiling point, to state the pressure. When the latter is 1 atm the term "normal boiling point" is employed.

During boiling bubbles of vapour are formed. That these bubbles do not collapse in upon themselves is a result of the equality between the pressure within the bubble and the pressure acting on the bubble. When the liquid is at a temperature such that its vapour pressure is less than that of the confining atmosphere, bubble formation is impossible.

The rate of evaporation is markedly increased when a liquid is brought to the boil. We have seen that the evaporation rate is dependent upon the amount of surface presented by the liquid. Bubble formation during boiling results in a considerable increase in the area of the liquid–vapour surface. This increase allows more molecules the opportunity of breaking free from the liquid surface.

5.8.   The Clapeyron equation

The quantitative relationship between vapour pressure and temperature is expressed by the Clapeyron equation:

$$\frac{dP}{dT} = \frac{L}{T \cdot \Delta V}$$

In this equation, the derivative dP/dT represents the rate at which the vapour pressure changes with temperature at the temperature $T(^\circ K)$. The quantity L is the latent heat of vaporization per mole, and for the transformation from liquid to vapour, is always positive. The quantity ΔV represents a volume change and for the transformation from liquid to vapour, is equal to the difference between the molar volumes of vapour and liquid,

i.e. $$\Delta V = V_{vapour} - V_{liquid}$$

Now at temperatures less than the critical value

$$V_{vapour} > V_{liquid}$$

and as a result ΔV, like L, will always have a positive value.

If the change being considered is one from vapour to liquid, then L will have a negative value as a result of heat being in this case liberated in the transformation (Chapter 11). The sign of ΔV also changes for now

$$\Delta V = V_{\text{liquid}} - V_{\text{vapour}}$$

Thus, the sign of the derivative dP/dT is at all times positive irrespective of how one views the transformation. This deduction is consistent with the experimental observation that the vapour pressure of all liquids increases with an increase of temperature.

The manner in which vapour pressure changes with temperature is shown in Fig. 5.4(a). Points on the curve (called the *vapour pressure curve*) represent the conditions of pressure and temperature under which both liquid and vapour exist in equilibrium. It should be noted that the derivative dP/dT is given by the slope of the curve at any temperature T.

The Clapeyron equation is rarely used in the form given for the liquid–vapour equilibrium. A more useful equation can be developed if certain fairly justifiable assumptions are made. Thus, if the temperature is far removed from the critical value

$$\Delta V \approx V_{\text{vapour}}, \quad V_{\text{vapour}} \gg V_{\text{liquid}}$$

Further, if we suppose the vapour to behave ideally

$$V_{\text{vapour}} = \frac{RT}{P} \quad n = 1$$

Hence
$$\frac{dP}{dT} = \frac{LP}{RT^2}$$

$\therefore \qquad\qquad \dfrac{dP}{P} = \dfrac{L}{RT^2} \cdot dT$

$\therefore \qquad\qquad d \ln P = \dfrac{L}{RT^2} \cdot dT$

If we regard the latent heat of vaporization to be a constant, independent of temperature, then

$$\int_{P_1}^{P_2} d \ln P = \frac{L}{R} \int_{T_1}^{T_2} \frac{1}{T^2} \cdot dT$$

$$\therefore \qquad \ln\frac{P_2}{P_1} = \frac{L}{R}\left(\frac{T_2 - T_1}{T_1 T_2}\right)$$

$$\therefore \qquad \log\frac{P_2}{P_1} = \frac{L}{2\cdot303R}\left(\frac{T_2 - T_1}{T_1 T_2}\right)$$

This last equation is particularly useful for it allows us to calculate the vapour pressure of a liquid at a second temperature, knowing its vapour pressure at a given temperature. Furthermore, in view of the fact that a plot of vapour pressure against temperature is equivalent to a plot of external pressure against boiling point, it follows that the above equation can be used to calculate the boiling point of a liquid under a variety of pressures.

EXAMPLE 5.1. *It is desired to produce superheated steam at* 120°C. *Under what pressure must water be boiled to achieve this? Normal boiling point of water* = 100°C. *Latent heat of vaporization of water* = 9726 cal mole^{-1}.

Basically, the information we wish to obtain for the solution of this problem is the vapour pressure of water at 120°C; let P be this value. In order to obtain a value for P we must know the vapour pressure of water at some temperature other than 120°C. Now we are told that water has a normal boiling point of 100°C — this is the temperature at which boiling commences under a pressure of 1 atm or 760 mm Hg. It follows from this that at 100°C, the vapour pressure of water is 760 mm Hg.

Hence $\qquad \log\dfrac{P}{760} = \dfrac{9726}{2\cdot303 \times 1\cdot987}\left(\dfrac{393 - 373}{393 \times 373}\right)$[†]

$$\therefore \qquad P = 1482 \text{ mm Hg.}$$

Therefore to produce superheated steam at a temperature of 120°C, water must be boiled under a confining pressure of 1482 mm Hg.

[†]Note that the value of R used in the calculation is $1\cdot987$ cal deg^{-1} mole^{-1}. We are obliged to use this value if the latent heat is expressed in cal mole^{-1}.

The value of the Clapeyron equation lies in the fact that it can be applied to *any* phase change. Thus, for the *sublimation process*, i.e. the process whereby solid passes directly into vapour without the intermediate formation of liquid, the relevant equations are of the same form as those already discussed. In this case

$$L = \text{latent heat of sublimation per mole,}$$

$$\Delta V = V_{\text{vapour}} - V_{\text{solid}} \approx V_{\text{vapour}},$$

and dP/dT = rate at which the vapour pressure of the solid changes with respect to temperature and is at all times positive.

The dependence of vapour pressure on temperature is shown in Fig. 5.4(b). The points on the curve (called the *sublimation curve*) represent the conditions of pressure and temperature under which both solid and vapour exists in equilibrium.

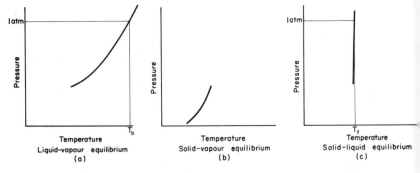

Fig. 5.4. Equilibrium conditions of pressure and temperature for liquid–vapour, solid–vapour and solid–liquid equilibria.

In the case of the solid to liquid transformation

$$L = \text{latent heat of fusion per mole,}$$

$$\Delta V = V_{\text{liquid}} - V_{\text{solid}},$$

and dT/dP = rate at which the freezing point changes with pressure.

Thus
$$\frac{dP}{dT} = \frac{L}{T(V_{\text{liquid}} - V_{\text{solid}})}$$

Now
$$L = lM; \; V_{\text{liq}} = v_{\text{liq}}.M; \; V_{\text{solid}} = v_{\text{solid}}.M$$

where l is the latent heat of fusion per gram, M is the molecular weight, and v_{liq} and v_{solid} are the *specific volumes* (cm^3g^{-1}) of the liquid and solid respectively.

∴
$$\frac{dP}{dT} = \frac{l}{T(v_{\text{liquid}} - v_{\text{solid}})}$$

Now for this particular change of state, ΔV is normally positive because most solids expand when melted. If then one views the transformation as taking place from solid to liquid, L, ΔV and consequently dP/dT are all positive, from which it follows that an increase in pressure elevates the freezing point as shown in Fig. 5.4(c). Points of this line (called the *freezing-point curve*) represent the conditions of pressure and temperature under which solid and liquid exist together in equilibrium.

A few substances, of which water is the prime example, experience a contraction upon melting, i.e. ΔV is negative, and this has the effect of making dP/dT negative. Consequently for such substances, the freezing point is depressed by an increase of pressure. Furthermore, because the molar (or specific) volumes of solid and liquid do not differ appreciably from each other, the quantity ΔV is small, thus making dP/dT large. This explains why the slope of the line in Fig. 5.4(c) is very nearly vertical, and indicates that a large change in pressure is necessary to produce even a small change in the freezing point. (The normal freezing point is the temperature at which solid and liquid are in equilibrium under a pressure of 1 atm.)

EXAMPLE 5.2. *Calculate the temperature at which ice melts under a pressure of* 11 *atm. The specific volumes of ice and water are* 1·0907 *and* 1·0001 *cm^3g^{-1} respectively at* 0°C *under* 1 *atm. The latent heat of fusion is* 80 *cal g*$^{-1}$.

Both l and ΔV are virtually constant and independent of pressure. Also, as we have already seen, the freezing point T, although changing with pressure, does so only slightly, so that we can write

$$\therefore \qquad \int_{P_1}^{P_2} dP = \frac{l}{T\Delta v} \int_{T_1}^{T_2} dT$$

$$\therefore \qquad P_2 - P_1 = \frac{l}{T\Delta v}(T_2 - T_1)$$

$$\therefore \qquad \frac{\Delta P}{\Delta T} = \frac{l}{T\Delta v}$$

or more conveniently for our purpose

$$\therefore \qquad \Delta T = \frac{T . \Delta v . P}{l}$$

This type of problem is one in which extreme care must be exercised in the choice of units in order that there should be numerical equality between both sides of the equation. The safest approach, although arithmetically tedious, is to express every quantity in absolute units. Thus

$\Delta v = v_{\text{liquid}} - v_{\text{solid}} = -0\cdot0906 \text{ cm}^3\text{g}^{-1}$
$T = 273°\text{K}$
$l = 3\cdot3472 \times 10^9 \text{ ergs g}^{-1}$ \qquad $1 \text{ cal} = 4\cdot184 \times 10^7 \text{ ergs}$
$\Delta P = 1\cdot013 \times 10^7 \text{ dynes cm}^{-2}$ \qquad $1 \text{ atm} = 1\cdot013 \times 10^6 \text{ dynes cm}^{-2}$

$$\therefore \atop \Delta T = \frac{-273 \text{ (deg)} \times 0\cdot0906 \text{ (cm}^3\text{g}^{-1}) \times 1\cdot013 \times 10^7 \text{ (dynes cm}^{-2})}{3\cdot3472 \times 10^9 \text{ (ergs g}^{-1})}$$

$$= -0\cdot075 \left(\frac{\text{deg dynes cm}}{\text{ergs}}\right)$$

$$= -0\cdot075 \text{ deg} \qquad\qquad 1 \text{ dyne cm} = 1 \text{ erg}$$

$\therefore$ Under a pressure of 11 atm, ice melts at $-0\cdot075°\text{C}$.

The three curves shown in Fig. 5.4 can be combined to give a composite diagram (Fig. 5.5).

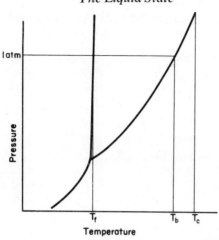

Fig. 5.5. Equilibrium conditions of pressure and temperature for liquid–vapour, solid–vapour and solid–liquid equilibria.

It will be noted that the three curves intersect at a point, referred to as the *triple point*, the significance of which is that it represents the *one* condition of pressure and temperature under which all three phases are in equilibrium with each other.

The critical temperature T_c is included to remind the reader that above this temperature the liquid cannot exist and so the vapour pressure curve has both an upper and a lower limit.

5.9. Refrigeration

Any refrigerator works on the principle that an evaporating liquid, through absorption of heat, cools its surroundings as well as itself.

An ideal refrigerant should have the following characteristics:

(i) It must be a gas which can be easily liquefied at ordinary temperatures.

(ii) It should have a high latent heat of vaporization.

(iii) It should be non-inflammable, non-toxic and non-corrosive.

Refrigeration on an industrial scale normally involves the use of ammonia as refrigerant, its principal advantage being that it has a higher latent heat than any other liquid except water. One such refrigeration unit is shown in Fig. 5.6. Here, ammonia is first compressed, then passed through condenser coils cooled by a water spray where the heat generated in the compression is removed. Also absorbed by the water is the heat of vaporization released in the condenser coils as vapour liquefies. The ammonia, now in the liquid condition, is forced through an

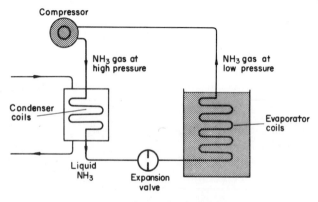

FIG. 5.6. Refrigeration unit.

expansion valve from which it emerges as a fine spray; evaporation then occurs in the evaporator coils, resulting in a cooling of the brine.† The ammonia vapour is then returned to the compressor and the cycle repeated. The cold brine may now be circulated to parts of an industrial plant where cooling is necessary at some stage in a manufacturing process, or where cold storage of either product or raw materials is essential. In a

---

†The "brine" is a calcium chloride solution which freezes at a temperature well below 0°C.

commercial ice plant, large tanks filled with water are placed in the brine and withdrawn when the water freezes.

In domestic refrigeration the same basic principles are employed involving the same basic components, namely the compressor, condenser, evaporator and expansion valve. However, a few modifications are necessary. Firstly, air-cooling is used to cool the compressed gas in the condenser. Secondly, ammonia is replaced by a less toxic refrigerant such as dichlorodifluoromethane which combines a fairly high value of latent heat of vaporization with a non-toxic nature.

5.10. The heat pump

Precisely the same components which go to make up a refrigerator are incorporated in what is referred to as a *heat pump*. The heat pump system differs from the refrigeration system in that the extracted heat, which in the latter is dissipated at the condenser, is in the former transferred from the condenser to a heating circuit, e.g. domestic central heating.

One such heat pump circuit is outlined in Fig. 5.7. From the liquid reservoir liquid at approximately room temperature passes through an expansion valve, emerging as a fine spray. This aids the vaporization which occurs under low pressure in the evaporator coils buried in the ground. Here the liquid, in addition to cooling itself, will, upon evaporating, absorb heat from its surroundings — in this case the cold ground. Emerging from the evaporator coils the vapour now at low pressure passes on to the compressor, where the high pressure is restored and the temperature is raised. Finally, in the condenser coils, the vapour condenses and in so doing releases its latent heat of vaporization. This, together with the heat generated in the compression stage, is then transferred to the central heating circuit. In this manner heat is transferred or "pumped" from outside to inside the house — hence the description "heat pump".

By itself this novel method of house-heating is not at present an economic proposition. However, by the simple inclusion of

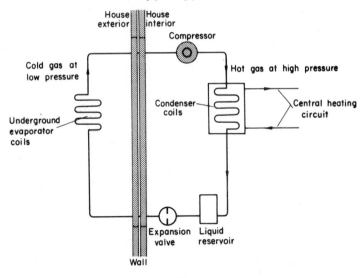

FIG. 5.7. Heat pump.

a few valves, the system becomes capable of functioning in a dual role, viz. house-heating in the winter – house-cooling in the summer. As the demand for summer comfort increases (i.e. in countries which experience hot summers) the system can be expected to become more economically viable.

In house-cooling, the system is made to function as a typical refrigerator. The change-over from heat pump to refrigerator is accomplished simply by reversing the flow of the gas. In this event, the outside coils become the condenser coils, whilst the inside coils become evaporator coils. In the latter, the liquid evaporates, absorbs its latent heat of vaporization from the surroundings (the house interior), then "pumps" this heat to the condenser. Here the vapour condenses and in the process loses its latent heat of vaporization to the ground.

CHAPTER 6

The Solid State

SOLIDS can exist in two states of aggregation, viz. *crystalline* and *amorphous* states. In the former state the basic units (molecules, atoms or ions) are arranged in space in an ordered fashion and because of this possess a high degree of symmetry. In the latter symmetry is less apparent because the units are arranged in a somewhat less regular pattern.

The manner in which the units arrange themselves in a crystal is largely determined by the nature of the forces acting between them. It is therefore convenient to classify solids in the following way:

Ionic solids — in which ions are attracted to each other by strong coulombic forces.

Covalent solids — in which atoms are strongly bonded to each other by covalent forces.

Molecular solids — in which molecules are bonded to each other by weak van der Waals forces or by dipole–dipole attractions as in, for example, hydrogen bonding.

Metallic solids — in which metallic ions are held together by the strong metallic bond.

These four classes we shall consider in turn.

6.1. Ionic solids

The most significant feature of the ionic bond is its non-directionality. Because of this, ions will tend to pack together as efficiently as possible. Thus, for sodium chloride (an AB salt type in which both ions carry the same charge) the most efficient close-packing of the ions is achieved when (i) each sodium ion is in contact with six chloride ions positioned at the corners of a regular octahedron, and (ii) each chloride ion is in contact with six sodium ions similarly positioned at the corners of a regular octahedron (Fig. 6.1).† The above configuration of ions is common to many inorganic compounds, e.g. the halides of

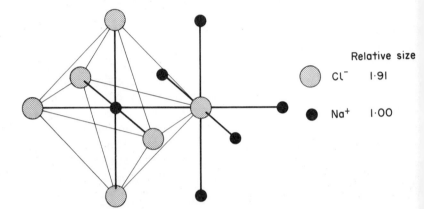

Relative size

Cl^- 1·91

Na^+ 1·00

FIG. 6.1. The sodium chloride structure.

lithium, sodium, potassium and rubidium. It does not, however, apply to the chloride, bromide and iodide of caesium, the fifth member of the alkali metal group of elements. The reason for this is that the ionic radius of the caesium ion is greater than the other alkali metal ions, so much so that it is comparable in size with the negative halide ions. Under these circumstances it becomes possible for each ion to pack around itself eight oppositely charged ions situated at the corners of a cube (Fig. 6.2).

†In discussing the structure of solids, some difficulty is encountered in describing the structures pictorially. This problem is most easily overcome by "exploding" the structure, i.e. separating the ions by distances which do not represent the true separation. This is done in Fig. 6.1 in the interest of clarity.

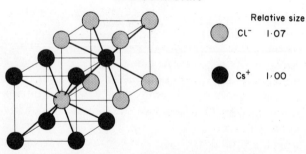

FIG. 6.2. The caesium chloride structure.

So far we have considered materials in which the charge on both the positive and negative ions has been the same. However, for salts of the AB_2 type this is not the case and the structure is accordingly more complex. A common structure for salts of this type is the fluorite (CaF_2) structure (Fig. 6.3). In this arrangement eight fluoride ions positioned at the corners of a cube are packed symmetrically around and in contact with the central calcium ion. The number of calcium ions surrounding any one fluoride ion is then predetermined at four because of the necessary condition of electrical neutrality. These four are situated at the corners of a regular tetrahedron, the calcium ion occupying the central position in the tetrahedron.

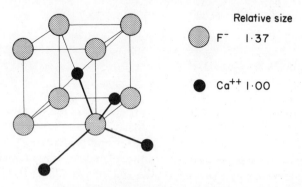

FIG. 6.3. The calcium fluoride structure.

6.2.   Covalent solids

A covalent bond is formed in a direction which gives maximum overlap of the electron clouds of the combining atoms (Section 1.7). The covalent bond is therefore a directed bond. In the case of carbon, four equivalent bonds are directed towards the corners of a regular tetrahedron. In combination with hydrogen, each sp^3 hybrid orbital of carbon overlaps with a hydrogen s orbital to produce methane. If now we imagine replacing each hydrogen atom in methane by a carbon atom, the resulting molecule C_5 has, unlike methane, unsatisfied valencies. These can become satisfied by the further addition of carbon atoms to produce a vast interlocking structure, a fragment of which is shown in Fig. 6.4. The above arrangement, showing the spatial relationship between carbon atoms

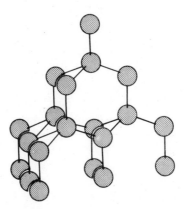

Fig. 6.4. The diamond structure.

in *diamond*, is, because of the strength and directional character of the covalent bond, a particularly rigid and hard structure — hence the use of diamonds in cutting and drilling tools. Similar structures are possessed by the elements silicon and germanium, and by the compounds zinc blende (ZnS) and silicon carbide (SiC).

Carbon, like a number of other elements (e.g. Sn, S, P), can have more than one crystalline form, a phenomenon referred to as *allotropy*. Different crystalline modifications of the same material are called *allotropes*; the second naturally occurring allotrope of carbon is *graphite*.

The structure of graphite is very different from that of diamond. It exhibits a layer structure in which each carbon atom is bonded to three others by sp^2 hybrid orbitals to form a planar network of hexagonal rings which can be thought of as being a fused aromatic ring system (Fig. 6.5). Whereas in benzene the

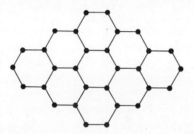

FIG. 6.5. Planar network of hexagonal rings in graphite.

nonhybridized p electrons are delocalized and free to move from carbon atom to carbon atom within the *ring*, so in the above fused ring system, the nonhybridized p electrons, being

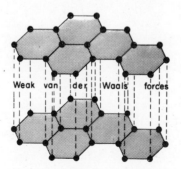

FIG. 6.6. The graphite structure.

similarly delocalized, have freedom of movement within the *layer*. This accounts for the fact that graphite is a conductor of electricity, whereas diamond, with all electrons localized about a particular bond, is an insulator (Section 6.4).

The layers in graphite are aligned parallel to each other and are bonded together by van der Waals forces (Fig. 6.6). Because such forces are weak, little effort is required to displace the layers relative to each other—hence the use of graphite as a high temperature lubricant (Section 28.3).

6.3. Molecular solids

If a molecular species is of a polar nature then dipole–dipole interactions are established between molecules. For example, in ice the positively charged hydrogen of one molecule assumes a position relative to the negatively charged oxygen in another molecule such that a net attraction results and a hydrogen bond forms. This leads to a structure which can best be appreciated by considering the spatial relationships between the oxygen atoms of different water molecules. Thus in Fig. 6.7, the central oxygen atom belonging to molecule 1 is surrounded tetrahedrally by four other oxygen atoms belonging to molecules 2, 3, 4 and 5. Because the hydrogen bond has, like the covalent

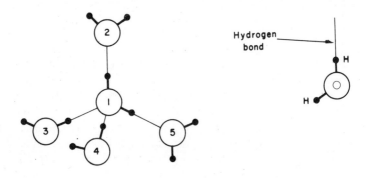

FIG. 6.7. The ice structure.

bond, directional characteristics, close packing of the water molecules is not achieved. Indeed the ice structure is very much an open one and this accounts for the fact that ice contracts rather than expands on melting. At the melting point some of the hydrogen bonds are broken, thus allowing the water molecules to pack more closely together.

For molecules having no polar character whatsoever (e.g. CO_2, Cl_2, O_2, N_2, C_6H_6) van der Waals forces act to hold the molecular units together and again, because these forces are particularly weak, such molecular solids have, in the main, low melting and boiling points.

Van der Waals forces, like coulombic forces, are non-directed and the way in which the molecules pack together is largely determined by their shape. This is reflected in the structure assumed by fairly simple molecules as compared with more complex ones in which the molecular shape is unsuited to efficient packing.

6.4. Metallic solids

Properties such as malleability, ductility, and good electrical and thermal conductivity are highly characteristic of metallic solids. These properties are a consequence of (a) the metallic bond and (b) the metallic structure.

(a) *The metallic bond*

None of the bonds so far considered can account for the fact that the atoms in a metallic crystal are very obviously attracted to each other by strong cohesive forces. To account for this it is necessary to introduce a new cohesive force called the *metallic bond*. This bond can be thought of as being similar to a covalent bond, in that there is a sharing of electrons. Whereas, however, in a true covalent bond the valency electrons are shared between two atoms, in the metallic bond the valency electrons can be thought of as being shared by all other atoms in the metal. It is this electron sharing that is responsible for the strong attractive forces in the metallic bond. We can therefore picture the situation in metals as follows: each atom looses control of its valency electrons and in so doing becomes a positive ion. The released electrons are mobile and are free to move throughout the entire structure of the metal. Thus, at any one point in time, one valency electron can be shared by two positive ions, while at some later point, it will be shared by other positive ions elsewhere in the structure.

The mobility of the valency electrons satisfactorily explains both the high thermal and electrical conductivity of metals. In a solid heat is transferred from one vibrating molecule to the next in a chain process. At high temperatures, the molecules, atoms or ions will vibrate more vigorously than at lower temperatures. The ease with which energy is transferred from one vibrating molecule to the next depends on the degree of perfection of the crystal structure and is strongly influenced by any structural imperfections such as the presence of impurity atoms.

Although the metallic structure is a particularly efficient one, it is by no means unique; the crystal structure of argon is similar to that of many metals yet its thermal conductivity is not to be compared with that of metals. Clearly some other factor must operate to make metals unique as far as this property is concerned. We can now appreciate what the nature of this other factor is, for in a metal energy can be transferred from one part of a metal to another through the agency of the mobile electrons in much the same sort of way that gas molecules transport energy, viz. through collision.

Electrical conductivity has much in common with thermal conductivity. The ability of metals to conduct a current is due to the fact that the mobile electrons, moving about quite randomly, will acquire a preference for moving in one direction only if placed under the influence of an applied voltage. (Ionic, covalent and molecular solids do not possess such mobile electrons and are therefore non-conductors, i.e. insulators.) The electrons now moving preferentially in one direction are, in their journey, obstructed by imperfections in the crystal structure and by the thermal vibrations of the metallic ions. Such vibrations become more and more energetic as the temperature rises and it is therefore hardly surprising that the electrical resistance should increase with increase of temperature. Conversely as the temperature is decreased, so also is the resistance. Most metals show electrical resistance even at temperatures approaching absolute zero—some, however, behave quite remarkably in this region of temperature, when their resistance disappears completely. In this condition the metals are said to be *superconductors*.

(b) *The metallic structure*

Most metals crystallize in one of three forms. These are:

- (i) the face-centred cubic close-packed (*fcp*) structure (Fig. 6.8(a)), e.g. Cu, Ag, Au, Pt.
- (ii) the hexagonal close-packed (*hcp*) structure (Fig. 6.8(b)), e.g. Cr, Mo, W, Ti.
- (iii) the body-centred cubic (*bcc*) structure (Fig. 6.8(c)), e.g. Li, Na, K, Cs.

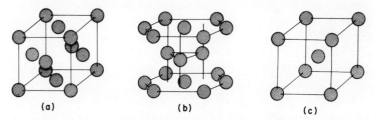

(a) (b) (c)

Fig. 6.8. Face-centred cubic, hexagonal and body-centred close-packed structures.

Because the metallic bond, like the ionic bond, is non-directional, the structure assumed by metals is principally determined by the various ways in which the spherical ions can pack together. The *fcp* and *hcp* structures are of special interest in this respect, for they represent the two ways in which equally sized spheres can be packed together in the most efficient manner, viz. one which makes the volume occupied by a large number of spheres a minimum. In both these structures the spheres occupy about 74 per cent of the total volume.

Let us now consider how these structures arise. First the spheres are arranged in layers (Fig. 6.9) in such a way that each sphere is in contact with six others. The second layer is then

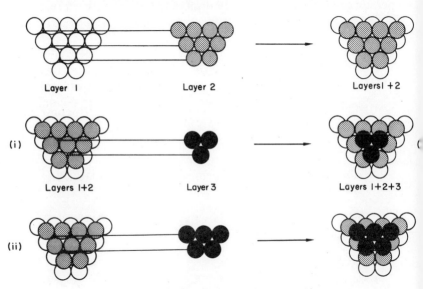

Layer I   Layer 2   Layers I + 2

(i)

Layers I+2   Layer 3   Layers I+2+3

(ii)

FIG. 6.9. Hexagonal and cubic close-packing.

placed on top of the first in such a way that all the second layer spheres lie in the cavities formed by the first layer of spheres. When a third layer is built up two possibilities exist, depending on which cavities are used in the second layer of spheres to accommodate the spheres of layer 3. Thus, in one case, third-layer spheres can be placed in second-layer cavities in such a way that (i) each sphere of layer three is directly above one in layer one to yield a hexagonal close-packed structure (Fig. 6.9(a)), or (ii) the configuration of third-layer spheres relative to the spheres of layers one and two is as shown in Fig. 6.9(b). This latter arrangement corresponds to a face-centred cubic close-packed structure.

One important difference between the *fcp* and *hcp* metals is that in the former, the metals are both malleable and ductile (i.e. they can be either hammered into thin sheets or drawn out into a wire), while in the latter, the metals tend to be both hard and brittle. This is accounted for by the fact that planes of ions can more easily glide over each other in the *fcp* structure than is the case in the *hcp* structure.

In the body-centred cubic structure, the spheres, occupying about 68 per cent of the total volume, are arranged in a less efficient manner, in that each sphere in a layer has four rather than six nearest neighbours (Fig. 6.10). As before, second-layer spheres fall into the cavities formed by the first-layer spheres. The addition of a third layer presents no problem in

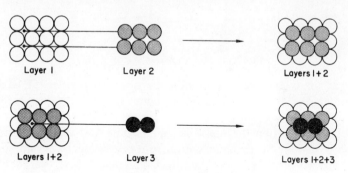

FIG. 6.10. Body-centred close-packing.

this case for in the third layer, each sphere must occupy a cavity which places it directly above one in layer one.

The alkali metals crystallize in this form and their relative softness is attributed to the more open (i.e. less close-packed) structure.

6.5. Imperfections in metallic crystals

There are basically two types of imperfections in metallic crystals. In one, the imperfection is of a physical character and can be introduced by imposing some strain on the metal. This creates a less ordered structure while at the same time altering the properties of the metal. Thus when a metal is *cold-worked* (i.e. hammered, rolled or pulled in the cold) it breaks up into small irregularly shaped crystals called *crystallites* which are oriented with respect to each other in a manner reminiscent of a mosaic pattern (Fig. 6.11). Within each crystallite the atoms are arranged in a regular fashion (viz. close-packed or body-centred) but as a result of the irregular

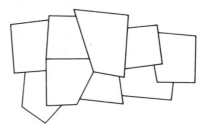

FIG. 6.11. Mosaic pattern of crystallites.

way in which the crystallites are oriented relative to each other, the susceptibility of the metal to flow is reduced, for although planes of atoms can glide over each other within a crystallite when the metal is under some sort of strain, this flow cannot be communicated from crystallite to crystallite. Thus cold-working has the effect of strengthening the metal, at the same time making it less ductile.

The reversal of the above process can be accomplished by a heat treatment referred to as *annealing*. Here thermal energy is sufficient to cause the mosaic structure to return to the more regular structure of a large crystal, thus reducing the strength of the metal and making it more ductile.

The second type of imperfection is of a chemical nature. In

this case, the structural regularity of the pure metal is disturbed by the presence of "foreign" atoms of either a metallic or a non-metallic element. The resulting substance is called an *alloy*.

There are two ways in which a "foreign" atom can be accommodated into the structure of a metal. In one, the "foreign" atoms enter at random the spaces or interstices between the "host" atoms to yield what are called *interstitial alloys*. In the second case, some of the "host" atoms are replaced by the "foreign" atoms to give *substitutional alloys*.

(a) *Interstitial alloys*

For an alloy of this type to be formed, the "foreign" atoms must be small. The elements hydrogen, boron, nitrogen and carbon are sufficiently small to be accommodated in this way. The inclusion of carbon into the structure of iron is of particular interest for the resulting alloy, *steel*, is the most important of all the alloys. (Metals such as chromium, tungsten, manganese, and molybdenum are also present in some steels depending on the desired properties of the steel.) Thus when carbon is added to molten iron it dissolves and on cooling, the melt crystallizes to an iron-carbon alloy with the carbon atoms occupying some of the interstices in the iron structure. As can be seen from Fig. 6.12, some distortion of the structure is inevitable and this has the effect of reducing the flow tendency. In turn this makes the iron less ductile but very much stronger.

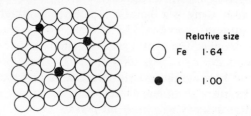

Relative size

◯ Fe 1·64

● C 1·00

FIG. 6.12. Iron–carbon alloy.

There is obviously a limit on the extent to which a "foreign" atom can be accommodated into the "host" structure – it is in fact reached before all the available spaces become occupied. In this condition the alloy will have a definite composition but should not be thought of as a true compound.

(b) *Substitutional alloys*

This type of alloy is formed when "host" and "foreign" atoms are of comparable size. Thus when a copper–nickel melt is cooled, a substitutional alloy is formed in which the nickel atoms are statistically distributed throughout the structure of the copper (Fig. 6.13). In this particular case the atomic sizes

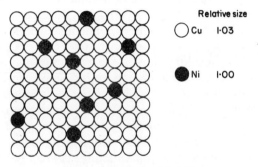

FIG. 6.13. Copper–nickel alloy.

of the two atoms are so similar that a substitutional alloy of any composition can be produced. This, however, is rather rare – usually there is a limit on the degree to which one atom can substitute for another. This situation arises when the difference in atomic sizes is in excess of about 15 per cent. Thus silver atoms can be substituted into the structure of copper (or vice versa) to a limited extent only, because of the distortion created by the substituent atoms (Fig. 6.14). As before, the distorted structures will have properties quite different from those of the parent structure.

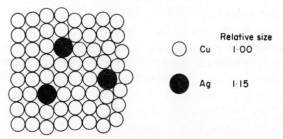

FIG. 6.14. Copper–silver alloy.

Up till now we have not mentioned the effect of alloying on the thermal and electrical conductivity of the metal. However, it should be clear that the presence of "foreign" atoms constitutes an imperfection which will therefore cause both conductivities to fall. Thus the thermal and electrical conductivities of copper are reduced considerably when it is alloyed with zinc (as in brass) or tin (as in bronze). Stainless steel, which because of the interstitial carbon has exceptional mechanical strength, exhibits very poor thermal and electrical conductivity for the same reason.

There are two other alloy types which cannot be discussed under the heading of crystal imperfections. In one type, the two metals form what are called *intermetallic compounds*. The formation of these compounds often fails to obey the normal valency rules. A lengthy discussion of such alloys is beyond the scope of this book. The remaining alloy type is one in which the alloy is simply an intimate mixture of the two metals. Each metal still retains its own identity and because of this the properties of the alloy are not appreciably different from those of the parent metal and in the main such alloys are of little industrial importance.

6.6. Semiconductors

Elements like boron, silicon, germanium and selenium have properties which do not allow them to be placed definitely in either the metallic or non-metallic group. They possess the typical metallic lustre of metals but exhibit very poor electrical conductivity. In conductivity terms their behaviour is intermediate between, on the one hand, metals (i.e. conductors) and on the other hand, non-metals (i.e. insulators) and for this reason they are called *semiconductors*.

One very surprising feature of semiconductors is that an increase in temperature brings about an increase in conductivity. The explanation for this is that the valency electrons are, at normal temperatures, almost exclusively associated with atom pairs (as in diamond) and are therefore immobile and incapable of transporting electricity. However, as the temperature is raised a number of valency electrons break away from their restrictive environment and become mobile — indeed the greater the thermal energy, the greater is their number. (That this situation does not arise in diamond is due to the firmness with which the valency electrons are shared — in silicon these electrons are less effectively localized.) As in the case of a metal, a temperature increase will cause the atoms of the semiconductor to vibrate with more vigour. This by itself would cause the conductivity to decrease. However, the superimposition of these two effects leads to an overall increase in conductivity with increasing temperature. This behaviour makes semiconductors particularly useful in the role of high precision thermometers where their high temperature coefficient of resistance makes them much more sensitive than normal resistance thermometers.

The valency electrons in a semiconductor can also be released through the absorption of radiant energy of a sufficiently low wavelength (i.e. high energy). When the conductivity is improved in this way the effect is called *photoconductivity*. This phenomenon finds many applications in, for example, light meters, infra-red detectors and photocopying.

The conductivity of semiconductors like silicon and germanium can be greatly enhanced by forming a substitutional alloy of these elements with the group V elements phosphorus, arsenic and antimony, the latter being present to the extent of only a few parts per million. Four of the five valency electrons of these impurity elements can be considered to be localized, forming four tetrahedrally directed covalent bonds. The remaining fifth electron cannot be so accommodated and being delocalized is free to act as a conducting electron (Fig. 6.15(a)). This type of alloy is called an *n*-type semiconductor (*n* for negative).

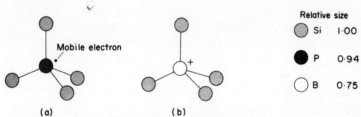

FIG. 6.15. *n*-type and *p*-type semiconductors.

A similar change in the properties of silicon or germanium can be accomplished by introducing trace amounts of the group III elements boron and aluminium to produce what is called a *p*-type semiconductor (*p* for positive). In this case, the impurity element has a deficiency of electrons and there is therefore introduced into the structure a "positive hole" which can be considered to be capable of migrating from one site to another under the influence of an applied voltage (Fig. 6.15(b)). It should not be thought that the two alloys considered possess any resultant charge – they are both quite neutral. The presence of negative and positive sites in the crystals follows as a consequence of the fact that the impurity elements are unable to satisfy the normal valency requirements in the "host" structure.

A most unusual effect is observed when an *n*-type and a *p*-type semiconductor are combined. The combination can be accomplished by "doping" one side of a wafer of very pure silicon or germanium with a group III element thus creating a *p*-type surface; an *n*-type surface is then built up by "doping" the other side of the wafer with a group V element. Now when this *n–p* combination is placed in an electrical circuit it is found that the resistance of this "cell" is direction dependent, being low when the current is moving in one direction, but very high when its flow is reversed. In other words, the *n–p* "cell" has the rectifying properties of a diode valve. Thus, if the polarity is as shown in Fig. 6.16(a), electrons will flow towards that part of the system which is electron deficient, i.e. from *n* to *p*. However, if the polarities are reversed (Fig. 6.16(b)), the electrons will tend to move towards that part of the system which is associated with an excess of electrons, i.e. from *p* to *n*. This unnatural tendency is resisted strongly with the result that only a very small current leak can be detected.

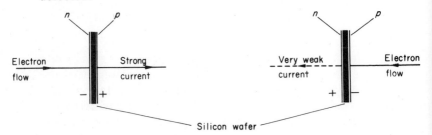

Fig. 6.16. The *n–p* combination.

Just as an *n–p* combination of semiconductors has the rectifying properties of a diode, so an *n–p–n* or *p–n–p* combination has both the rectifying and amplifying properties of a triode. Such a combination is known as a transistor, the development of which has revolutionized the electronics industry.

CHAPTER 7

Cements

ALTHOUGH the term 'cement' may be broadly used to describe any adhesive material, it is more generally used to describe the lime-based products used in civil engineering to bind together coarse aggregates, and it is this type of cement we shall discuss here. In particular we shall concern ourselves with *hydraulic cements*, i.e. cements which are capable of setting and hardening under water due to the chemical interactions between water and the constituents of the cement.

7.1. Portland cement

By far the most important cement is *Portland cement*, which is the essential binding material in concrete. This is manufactured by igniting a mixture of limestone (calcium carbonate) and clay (alumino-silicates) in a rotary kiln at temperatures up to 1500°C. At these high temperatures partial fusion of the material occurs, and on cooling *cement clinker* is obtained. This is mixed with gypsum (calcium sulphate) and ground to a fine powder.

The material obtained after the roasting process is fundamentally different from the mixture fed into the kiln. At the high temperature used, water and carbon dioxide are driven off and the solid material remaining reacts to form a variety of chemical compounds, depending on the exact composition of the original mixture. Portland cement contains four main constituents, tricalcium silicate, β-dicalcium silicate, tri-calcium aluminate and a ferrite phase which is a solid solution of somewhat

103

variable composition. The formulae of these materials are shown in Table 7.1.

TABLE 7.1. PRINCIPAL CONSTITUENTS OF PORTLAND
CEMENT

Constituent	Formula	Abbreviated formula
Tricalcium silicate	$3CaO \cdot SiO_2$	C_3S
β-Dicalcium silicate	$2CaO \cdot SiO_2$	C_2S
Tricalcium aluminate	$3CaO \cdot Al_2O_3$	C_3A
Ferrite phase	$4CaO \cdot Al_2O_3 \cdot Fe_2O_3$	$C_4AF.$

Two points are worthy of note at this stage. Although the formulae of the compounds listed in the above table are written as mixed oxides, it must not be inferred that these oxides exist as such in the actual compound. The first compound listed, tricalcium silicate, could equally well be represented by the formula Ca_3SiO_5. The formula simply gives the information that the compounds in question contain certain elements in fixed proportions. The mixed oxide formulae are normally further abbreviated by the cement chemist as shown in the final column of Table 7.1, where each oxide is represented by a single letter, e.g.

$$CaO \quad SiO_2 \quad Al_2O_3 \quad Fe_2O_3 \quad H_2O$$

$$C \quad \quad S \quad \quad A \quad \quad F \quad \quad H$$

7.2. Hydration reactions

It has already been stated that cementing action arises from the chemical interaction of water with the compounds present in the original cement. The four principal constituents as formed in the roasting process are anhydrous materials which are unstable in the presence of water. If therefore water is added to these materials, the system is not in a state of equilibrium and chemical reaction must occur.

Tricalcium silicate

When finely ground tricalcium silicate is added to water a fairly rapid hydration reaction occurs with the formation of a gelatinous hydrated calcium silicate and crystals of calcium hydroxide. The formation of the hydrated silicate on the surface of the anhydrous particles tends to slow the hydration reaction considerably. A suggested equation is

$$2Ca_3SiO_5 + 6H_2O \rightarrow Ca_3Si_2O_7.3H_2O + 3Ca(OH)_2$$

β-Dicalcium silicate

This ingredient reacts very slowly with water, possibly because it has a more closely packed structure than the tricalcium silicate. Again, the products are lime and a hydrated silicate which analyses $Ca_{3.3}Si_2O_{7.3} \cdot 3 \cdot 3H_2O$. Although the two hydrated products are not chemically identical their physical structures are remarkably similar.

Tricalcium aluminate

The reaction of this material with water is rapid and is accompanied by a strong evolution of heat. The reaction differs from the previous ones in that no lime is produced, the main product being the crystalline hydrate $3CaO \cdot Al_2O_3 6H_2O$.

Ferrite

As with the previous material reaction with water is rapid, the products including the compounds $3CaO \cdot Al_2O_3 \cdot 6H_2O$ and $3CaO \cdot Fe_2O_3 6H_2O$.

The development of compressive strength on hydration for each of the individual compounds is shown in Fig. 7.1, where it can be seen that tricalcium silicate develops a high strength extremely rapidly, dicalcium silicate rather more slowly, and the remaining materials never develop a comparable strength.

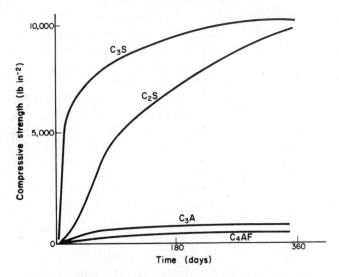

FIG. 7.1. Development of compressive strength of cement compounds. (Reprinted from Bogue and Lerch, *Ind. Eng. Chem.* **26** (1934) 837, by permission of the American Chemical Society)

If we now consider the addition of water to cement itself, i.e. to a ground mixture of the four main constituents, we find that for a useful cement the rates of the individual hydration reactions become extremely important. We must at this point distinguish between *setting* and *hardening*. Setting is the initial stiffening of the material and normally occurs within a few hours of mixing, while hardening is the slow development of strength which proceeds over a number of years.

When mixed with water, simple cement paste hydrates rapidly with a strong evolution of heat – a process described as *flash set*. Since a cement paste must in practice remain workable over a period of time, gypsum is added to the main ingredients and retards the setting process. The phenomenon of flash set is generally attributed to the fast hydrolysis of tricalcium aluminates to form insoluble complexes of the type

$3CaO \cdot Al_2O_3 \cdot 3CaSO_4 \cdot 31H_2O$. The retarding action is probably due to the formation of insoluble coatings on the highly reactive tricalcium aluminate surface, this slowing the hydrolysis reactions. The mechanism is not simple, however, and in the retarding of Portland cement it is desirable to have both gypsum and lime present, since a mixture of these two compounds is far more effective than either acting alone.

Any inorganic calcium salt can in fact act as a retarder in the hydrolysis of Portland cement but some (e.g. $CaCl_2$, $Ca(NO_3)_2$), while retarders at low concentrations, actually accelerate the setting reaction when present in higher concentrations.

7.3. Hardening of Portland cements

From what we have already learned it is obvious that the reactions which can occur when cement powder is mixed with water are both numerous and complex. It is hardly surprising therefore that controversy arose as to the actual mechanism of the cementing action.

One theory, that of crystallization, arose because the setting of plaster of Paris could be attributed to a rapid crystallization process which yielded a cohesive mass of small crystals. Plaster of Paris $(CaSO_4 \cdot \frac{1}{2}H_2O)$ is much more soluble than gypsum $(CaSO_4 \cdot 2H_2O)$, so that when the hemihydrate is mixed with water and becomes hydrated, the solution is super-saturated with respect to the dihydrate. A rapid crystallization results, with the formation of an interlocking mass of very small needle–like crystals to yield a solid of considerable strength. It was believed that a similar situation could arise in the setting of Portland cement, whereby the cement ingredients could dissolve to form a solution super-saturated with respect to the hydration products. Rapid crystallization should then result to give a highly cohesive mass of small crystals.

An alternative view was that the hardening of cement paste was due to the formation of a *gel*, which initially would be grossly swollen with water, but would eventually harden to a solid mass. A gel is a colloidal system of at least two components and mechanically generally behaves as a typical solid. The dispersed component may be either crystalline or amorphous, but forms an irregular open framework within which the dispersion medium is contained, as illustrated in Fig. 7.2.

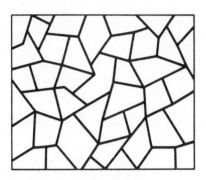

Fig. 7.2. Gel structure.

Modern theory amalgamates these two opposing views. The initial reaction does appear to involve solution of the cement, followed by precipitation of hydrated products. Although these complex hydrates may be described as crystalline, the particles are of colloidal size and the structures are normally designated as gels.

Figure 7.3 shows schematically how the anhydrous cement powder may be converted into a hard solid mass. Figure 7.3(a) shows the initial situation with discrete cement grains dispersed in water. Chemical interaction occurs to give a supersaturated solution from which hydrated products precipitate at the surface of the cement grains as shown in Fig. 7.3(b). This process continues with the formation of a gel, both inwards into the original cement grains and outwards into the voids between the grains, until the products around each grain begin to merge (Fig. 7.3(c)). This is the setting stage. In Fig. 7.3(d) it can be seen that hardening has occurred, and the

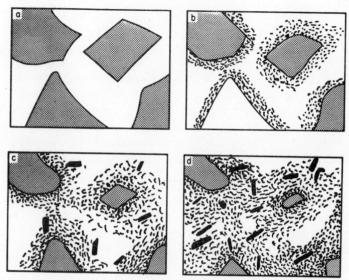

FIG. 7.3. Mechanism of cementing action (Reprinted from *The Chemistry of Cements*, Vol. 1, H. F. W. Taylor; Academic Press, 1964).

spaces between grains have become progressively more closely packed. Some relatively large crystals begin to form at the setting stage, and more develop by recrystallization processes.

7.4. Weathering of cements

Since concrete made with Portland cement is so widely used as a structural material, some knowledge of its resistance to chemical attack is essential. Even pure water is capable of decomposing cement given the correct conditions. The principal mechanism is the leaching of lime from the set cement, leaving eventually a soft spongy structure with no cohesive strength. Fortunately this leaching action is extremely slow unless the water can flow continuously through the cement material, and deterioration is generally negligible.

The action of acidic waters on cement structures is rather more serious, deterioration again being caused by a dissolution of some of the constituents. Acidity in natural waters usually arises from dissolved carbon dioxide or humic acid derived from decayed vegetable matter, the latter not being particularly active. The outer surfaces of hardened cement contain calcium carbonate formed by the action of atmospheric carbon dioxide on the lime in the cement, and this provides some degree of protection against attack by acidic waters. The carbonate can be dissolved, however, by conversion into bicarbonate:

$$CaCO_3 + H_2CO_3 \rightarrow Ca(HCO_3)_2$$

Since the permeability of the set cement is an important factor in water resistance, one obvious method of protection is to render the cement less permeable to water, either by internal or external treatment. Materials added to the cement mix include inert materials which simply act as pore fillers, e.g. clays, ground silica; inert materials which act as water repellants, e.g. waxes, mineral oils and bitumen; and active water repellants which actually react chemically with the cement, e.g. sodium or potassium soaps which react to form insoluble calcium soaps within the cement. It is possible that these internal additives act simply as workability aids, thus giving a more uniform and therefore stronger end product.

Surface treatments include drying oils, bituminous paints and even complete lining by sheet plastics or glass may be used when extremely corrosive materials are present, e.g. in certain industrial applications. These coatings successfully stop deterioration for a time, but since they have only limited resistance to abrasion, periodic renewal of the protective films is necessary.

One of the most serious causes of deterioration of cement structures is the presence of sulphates which can react both with the lime and the hydrated calcium aluminates in the cement, e.g.

$$Ca(OH)_2 + Na_2SO_4 \rightarrow CaSO_4 \cdot 2H_2O + 2NaOH.$$

Magnesium sulphate is particularly corrosive in that it attacks the hydrated calcium silicates far more readily than do sodium or potassium sulphates. Apart from the fact that the products of sulphate attack have less binding power than the original materials, they generally occupy a much greater solid volume, thus local expansion and disruption of the mortar occurs.

Deterioration due to sulphate-containing waters is found not only in certain soils, but also where cement is exposed to sea water, which contains considerable amounts of magnesium sulphate as well as a high concentration of chlorides. Again the initial chemical attack is the formation of calcium sulphate and sulphoaluminates, but whereas with simple sulphate waters deterioration is caused by expansion of the mortar, in sea water no expansion is found. Instead, the structure is weakened by a severe leaching process due to the increased solubility of the reaction products in sea water and to the continuous agitation by wave motion.

Other mechanisms operate concurrently with sulphate attack, e.g. the cement structure is disrupted by the crystallization of salts within the pores due to alternate wetting and drying by waters of high salt content. Reinforced concrete is particularly susceptible to attack by sea water, since corrosion of the reinforcing metal causes further cracking of the structure, thus accelerating the deterioration.

One method of increasing the sulphate resistance of Portland cement is to increase the amount of the ferrite phase (C_4AF) at the expense of the calcium aluminate (C_3A). Alternatively a different type of cement such as alumina cement (Section 7.5) may be used where sulphate waters are present. A further method is to cure the cement in steam at temperatures greater than 100°C. This process brings about several changes in the cement constituents, e.g. the free lime forms calcium silicates, while the tricalcium and dicalcium silicates react to form more stable crystalline hydrates, as does tricalcium aluminate. The overall effect is therefore to produce a cement much more resistant to sulphate attack.

7.5. Special purpose cements

Alumina cement

This is manufactured by roasting limestone and bauxite $(Al_2O_3 \cdot xH_2O)$ to give a product containing about equal amounts of alumina and lime, together with a substantial amount of iron and a small percentage of silica. The principal constituent after roasting is monocalcium aluminate (CA), the iron occurring as a ferrite solid solution of composition between C_4AF and C_6AF_2. The hydration reactions are similar to those occurring in Portland cement, the main hydrate being the compound CAH_{10}, although other hydrates may be produced if the temperature is greater than 25°C. This material is far more crystalline than Portland cement, but still the crystal particles are extremely small.

Although there is little difference in the setting time of alumina cement as compared to that of Portland cement, the former thereafter develops high strength very rapidly, nearing its maximum strength even after only 24 hours. As already described alumina cement has a high resistance to sulphate attack.

Another feature of alumina cement is its use in refractory concrete. The concrete is mixed in the normal way, but the aggregate itself must have heat-resisting properties, e.g. crushed firebrick. When the concrete is heated it gradually loses strength until a minimum is reached at about 900°C. At temperatures greater than this strength again develops due to the formation of new bonds between the cement products and the aggregate. Concretes based on alumina cements together with pure alumina aggregate can be used up to 1800°C. These high-temperature cements are widely used in the manufacture of kilns and furnaces.

White and coloured cements

Ordinary Portland cement is grey because of its iron content and for a white cement this must be kept as low as possible. The raw materials must therefore be a suitable limestone, e.g. chalk, and china clay, both of very low iron content. The ultimate strength of such a cement is generally lower than normal Portland cement.

Although Portland cement can be coloured by pigments, a wide range of colours is not practicable because the original grey colour tends to persist. Instead, white cement with an appropriate pigment is generally used. The achievement of lasting colour is difficult, not because of any defect in the pigment itself, but because of the development of a white calcium carbonate film over the surfaces. This can be removed by treatment with dilute mineral acids, but the treatment must be carefully controlled due to the possible attack of the acid on the cement itself.

Expansive cements

Normal cements undergo a contraction in volume on drying, and this causes obvious manipulative difficulties. Although no cement has yet been discovered which dries without any volume change, some progress in this direction has been made, mainly by using additives to Portland cement. The aim has been to produce a cement which expands initially, the expansion counteracting the final shrinkage on drying, so that no net volume change occurs.

By igniting a mixture of bauxite, limestone and gypsum, a calcium sulphoaluminate is obtained which when ground with Portland cement clinker gives the desired properties. The material is not particularly easy to use and careful control of the curing process is essential.

Low heat cements

The hydration of the anhydrous compounds of normal Portland cement is an exothermic process and liberates a considerable quantity of heat. This can be of particular importance in constructions where large masses of concrete are involved, leading to large temperature gradients within the concrete mass. Temperature rises of up to 50°C have been observed. The principal danger is that the eventual cooling of the structure will be accompanied by crack formation, which could lead to early failure of the material.

The main sources of heat are the hydration reactions of tricalcium aluminate and silicate. The amount of the former can fairly readily be decreased without any detriment to strength by increasing the Fe/Al ratio thus converting the C_3A material to C_4AF. The effect of the silicate can be reduced by increasing the proportions of β-C_2S at the expense of C_3S, although this does reduce the early strength of the set cement (see Fig. 7.1).

Pozzolanic cements

In Portland cement anhydrous calcium silicates are prepared by roasting, and the subsequent hydration reactions give a material with binding properties. Pozzolanas are materials which, although not in themselves cements, will react with lime and water at normal temperature to produce a cement-like material. These materials when occurring naturally are usually of volcanic origin, but can be made artificially by roasting such materials as clays, shales and pulverized fuel ash.

Pozzolanic cements are made by grinding together pozzolanas with Portland cement to a fine powder, which is then treated as a normal cement. The Portland cement constituents appear to hydrate normally, but the lime produced in the hydration reactions further reacts with the pozzolanas in the presence of water.

These cements possess several desirable characteristics, e.g.

the heat of hydration is low, their resistance to chemical attack is very good, and their ultimate strength can be greater than that of conventional Portland cement. They do, however, suffer from the disadvantage that their rate of hardening is rather slow. At low temperatures this characteristic is so accentuated that in certain climates their use is not practicable.

CHAPTER 8

Electrochemistry and Metallic Corrosion

8.1. Oxidation and reduction

Although at a very elementary level it is convenient to define oxidation as simply the gain of oxygen or loss of hydrogen, such a definition has only limited application. If, for example, we wish to describe the conversion of ferrous chloride to ferric chloride as an oxidation reaction we must

$$FeCl_2 + \tfrac{1}{2}Cl_2 \rightarrow FeCl_3$$

obviously redefined the term oxidation. The most convenient definition is that *oxidation involves the loss of electrons, reduction involves the gain of electrons.*

If we re-examine the simple reaction above we find we can consider the overall process to be the sum of two parts. The ferrous iron is oxidized to ferric iron, involving the loss of one electron, while the oxidizing agent chlorine gains

$$Fe^{2+} \rightarrow Fe^{3+} + 1e$$

an electron and is reduced to the negatively charged chloride ion

$$\tfrac{1}{2}Cl_2 + 1e \rightarrow Cl^-$$

Obviously the two simple reactions, called *half-reactions*, are complementary and must occur in equivalent amounts, i.e. the electrons generated in the oxidation reaction must be consumed in the reduction.

If we now imagine the two half-reactions to occur in separate vessels, but joined by a suitable conductor, we can see that if the overall reaction still takes place, the transfer of electrons will occur through the conductor, i.e. an electric current will have been generated.

Electrochemistry is the study of systems such as this where chemical reactions are used to generate an electric current, i.e. *electrochemical cells*, together with the converse process where an applied current is used to bring about chemical change, i.e. *electrolysis*.

8.2. Standard electrode potentials

If a rod of metal, e.g. zinc, is immersed in water or a solution containing zinc ions, a potential difference is set up between the rod and the solution. Generally the tendency is for ions from the metal rod to go into solution, leaving the metal negatively charged:

$$Zn \rightleftharpoons Zn^{2+} + 2e$$

The size of the *electrode potential*, i.e. the potential between the rod and solution, is determined by the nature of the metal and the concentration of metal ions in the solution.

If we now consider a different metal, e.g. copper, we find that this rod is positively charged with respect to the solution. Here the reverse reaction has occurred, metal ions leaving the solution and being deposited on the metal surface where they acquire electrons

$$Cu^{2+} + 2e \rightleftharpoons Cu$$

If these two *half-cells* are left in isolation, equilibrium is rapidly attained and reaction ceases. They can, however, be coupled together as is shown in Fig. 8.1. The two half-cells are separated by a porous partition, and the two metal rods joined by a wire. The high concentration of electrons generated in the reaction no longer builds up on the zinc rod, because

$$Zn \rightleftharpoons Zn^{2+} + 2e$$

the electrons are now free to travel along the wire to the copper

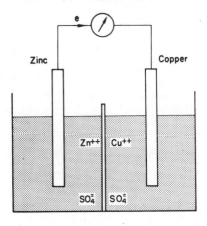

FIG. 8.1. The zinc–copper cell.

electrode where they are used in the reaction

$$Cu^{2+} + 2e \rightleftharpoons Cu$$

The overall reaction occurring in the cell is

$$Zn + Cu^{2+} \rightleftharpoons Cu + Zn^{2+}$$

i.e. the zinc is being oxidized and the cupric ion reduced.

The tendency for any given element to be oxidized or reduced in a galvanic cell can be measured by coupling together two half-cells, one of whose potentials is already known. By convention, the reference half-cell is the standard hydrogen electrode which is assigned a potential of zero. The hydrogen cell, illustrated in Fig. 8.2, consists of a platinum electrode coated with platinum black, partially immersed in a 1·0 molar hydrogen ion solution, while hydrogen gas at a pressure of one atmosphere is bubbled over the electrode surface. The standard reference temperature is 25°C.

We can now see that in order to determine the standard electrode potential of, say, zinc, we would connect a standard hydrogen cell to a standard zinc cell made by immersing a zinc

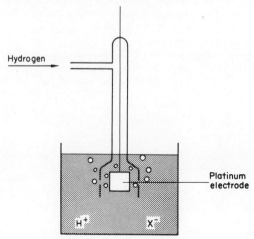

FIG. 8.2. The hydrogen electrode.

rod in a solution 1·0 molar in zinc ions at 25°C, and measure
the potential of the cell. This cell potential is found to be 0·76
volts, and the sign convention we shall use (though not universally accepted) is that the standard electrode potential of zinc
is negative, i.e. −0·76 volts.

On this basis standard electrode potentials of the majority
of elements yielding ions in solution have been obtained. These
elements can then be listed in order, forming the *electrochemical series*, as is shown in Table 8.1. This is an extremely
useful table since it lists elements according to their ease of
oxidation and can be used as a rough guide as to how certain
elements and ions will react. If we consider reactions involving
positive ions, any element will displace any other below it in
the series. As we have already seen, zinc displaces copper
ions:

$$Zn + Cu^{2+} \rightarrow Zn^{2+} + Cu$$

Similarly, for reactions involving negative ions, any element
will displace one above it in the series, e.g. chlorine will displace iodine:

$$\tfrac{1}{2}Cl_2 + I^- \rightarrow \tfrac{1}{2}I_2 + Cl^-$$

TABLE 8.1. STANDARD ELECTRODE POTENTIALS

Reaction	$E°$ (volts)	Reaction	$E°$ (volts)
$Li^+ + e \rightleftharpoons Li$	−3·05	$Pb^{2+} + 2e \rightleftharpoons Pb$	−0·13
$K^+ + e \rightleftharpoons K$	−2·93	$2H^+ + 2e \rightleftharpoons H_2$	0·00
$Ca^{2+} + 2e \rightleftharpoons Ca$	−2·87	$Cu^{2+} + 2e \rightleftharpoons Cu$	+0·34
$Na^+ + e \rightleftharpoons Na$	−2·71	$I_2 + 2e \rightleftharpoons 2I^-$	+0·54
$Mg^{2+} + 2e \rightleftharpoons Mg$	−2·37	$Ag^+ + e \rightleftharpoons Ag$	+0·80
$Al^{3+} + 3e \rightleftharpoons Al$	−1·66	$Br_2 + 2e \rightleftharpoons 2Br^-$	+1·07
$Zn^{2+} + 2e \rightleftharpoons Zn$	−0·76	$Cl_2 + 2e \rightleftharpoons 3Cl^-$	+1·36
$Fe^{2+} + 2e \rightleftharpoons Fe$	−0·44	$Au^{3+} + 3e \rightleftharpoons Au$	+1·50
$Ni^{2+} + 2e \rightleftharpoons Ni$	−0·25	$F_2 + 2e \rightleftharpoons 2F^-$	+2·65
$Sn^{2+} + 2e \rightleftharpoons Sn$	−0·14		

8.3. Non-standard conditions

In the previous section all measured electrode potentials referred to a standard state of 25°C and molar solutions of metal ions. Since we cannot expect that in practice metals will always be in contact with solutions which are 1·0 molar with respect to the appropriate ions, we must consider the effect of deviating from these idealized conditions.

For a metal electrode where the half-cell reaction is written

$$M^{n+} + ne \rightarrow M$$

the electrode potential for any concentration is given by the equation

$$E = E° + \frac{RT}{nF} \ln [M^{n+}]$$

where $E°$ is the standard electrode potential,

R is the gas constant,

T is the temperature (°K),

F is the Faraday (96,500 coulombs),

$[M^{n+}]$ is the concentration of M^{n+} (g ion litre^{-1})

At 25°C this equation simplifies to

$$E = E° + \frac{0·059}{n} \log [M^{n+}]$$

We can now consider the combination of two Cu/Cu^{2+} half-cells, as in Fig. 8.3 where the electrode is in each case a copper rod, but the solution of cupric ions is not the standard 1·0 molar in each half-cell. We discover that a galvanic cell has been produced, just as happened when a copper and a zinc rod were connected.

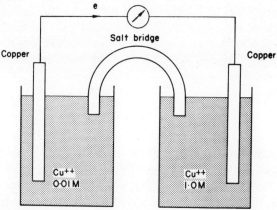

FIG. 8.3. A Cu/Cu^{2+} concentration cell.

The potential in one half-cell is

$$E_1 = E^\circ + \frac{0\cdot059}{2} \log 0\cdot01$$

while that in the other half-cell is

$$E_2 = E^\circ + \frac{0\cdot059}{2} \log 1\cdot00$$

The overall potential of the cell is the difference between the two half-cell potentials

$$E_2 - E_1 = \frac{0\cdot059}{2} \log\frac{1\cdot00}{0\cdot01}$$

$$= 0\cdot059 \text{ volts}$$

Thus a galvanic cell has been established such that copper will dissolve at electrode 1 and cupric ions will deposit at electrode 2 until the concentration of cupric ions is equalized in the two half-cells.

We have now learned what is meant by galvanic action, and how to predict the relative ease of oxidation of a given element. It must be emphasized, however, that although a study of electrode potentials may be a useful guide as to the feasibility of a reaction taking place, such a study tells us nothing about the rate at which the reaction proceeds. In certain cases the rate of a perfectly feasible chemical reaction may be so slow as to appear to us not to occur.

8.4. Electrolysis

Just as we have described an oxidation-reduction reaction in terms of two half-reactions of a galvanic cell, we can consider oxidation-reduction reactions to occur in electrolysis. Here, when direct electric current is passed through an electrolyte, cations migrate to the *cathode* (negative electrode) where they accept electrons and are reduced, while anions migrate to the *anode* (positive electrode) where they lose electrons and are therefore oxidized.

If we consider the electrolysis of molten sodium chloride the following processes occur:

(i) migration of ions to the electrodes
(ii) at the cathode

$$Na^+ + 1e \rightarrow Na$$

(iii) at the anode

$$Cl^- \rightarrow \tfrac{1}{2}Cl_2 + 1e$$

Thus the overall reaction may be written

$$Na^+ + Cl^- \rightarrow Na + \tfrac{1}{2}Cl_2.$$

In the above example the only ions present are those of sodium (Na^+) and chloride (Cl^-), and these inevitably must be the ions discharged. When more than one ion of given sign is present, one particular ion is always more readily discharged than the others. The ease of discharge of an ion is largely determined by the value of its *discharge potential*, which is the minimum potential which must be applied between the solution and the electrode before the ion is discharged. The numerical value of the discharge potential is usually approximately equal to the standard electrode potential.

A complication arises when discharge reactions involving the formation of gases are considered. Here the discharge potential may be much greater than the electrode potential, the difference in value being called the *over-voltage*. Its magnitude depends on several factors, the most important being the nature of the electrode.

An example of the significance of this is found in the common lead storage battery. This cell is composed of two electrodes, lead and lead dioxide, in contact with sulphuric acid. When supplying current, the lead is oxidized to lead ions (Pb^{2+}), while the lead dioxide is reduced to lead ions (Pb^{2+}). The half-reactions may be formally represented:

$$Pb \rightarrow Pb^{2+} + 2e$$
$$PbO_2 + 4H^+ + 2e \rightarrow Pb^{2+} + 2H_2O$$
$$\overline{PbO_2 + 4H^+ + Pb \rightarrow 2Pb^{2+} + 2H_2O}$$

i.e. $$Pb + PbO_2 + 2H_2SO_4 \rightarrow 2PbSO_4 + 2H_2O$$

In the discharge reaction, sulphuric acid is used up and lead sulphate is produced at each electrode.

If we now consider the reverse process of charging the battery, a potential is applied and electrolysis occurs. At the cathode two reactions are possible:

$$Pb^{2+} + 2e \rightarrow Pb$$
$$H^+ + 1e \rightarrow \tfrac{1}{2}H_2$$

The electrochemical series would suggest that the discharge of hydrogen would be the favoured reaction. However, the over-voltage of hydrogen on lead raises the discharge potential of hydrogen to a value greater than that of lead, and lead is in fact deposited at the electrode. Similarly at the anode, a possible reaction would be the evolution of oxygen:

$$2H_2O \rightarrow O_2 + 4H^+ + 4e$$

but again this does not occur because of the over-voltage effect and the reaction which is preferred is the oxidation of lead ions (Pb^{2+}) to lead dioxide.

The process of electrolysis has many technical applications, e.g. in metal refining which will be discussed in Chapter 9, in the preparation of certain inorganic and organic materials, and in electroplating where a base metal is coated with a thin layer of another suitable metal, e.g. silver or chromium, by electrolytic deposition.

8.5. Fuel cells

A different type of cell which is potentially of great importance is the *fuel cell*, which is a means of converting conventional fuels directly into electric current. The fuel is supplied to one electrode and an oxidant to the other.

Perhaps the simplest cell of this type is the hydrogen–oxygen cell, which was first successfully demonstrated more than 100 years ago. Here the 'fuel' is hydrogen gas which is supplied to the anode, while oxygen is supplied to the cathode as shown in Fig. 8.4. At the anode hydrogen is adsorbed and gives up electrons to the electrode, forming hydrogen ions. At the cathode, adsorbed oxygen accepts electrons to form hydroxyl ions:

Anode	Cathode
$2H_2 \rightarrow 4H$ (adsorbed)	$O_2 \rightarrow O_2$ (adsorbed)
$4H$ (adsorbed) $\rightarrow 4H^+ + 4e$	O_2 (adsorbed) $+ 2H_2O + 4e \rightarrow 4OH$

$$H^+ + OH^- \rightarrow H_2O$$

The reaction product is water, just as in the normal combustion of hydrogen.

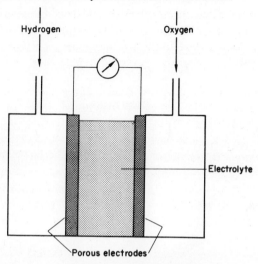

FIG. 8.4. A H_2/O_2 fuel cell.

Gaseous fuels require porous electrodes so that the reaction takes place at a gas–liquid–solid interface. Gaseous fuels other than hydrogen which have been successfully used include the readily available hydrocarbons methane, ethane and propane. In these cases the reaction products include carbon dioxide as well as water; e.g. at the anode:

$$C_3H_8 + 6H_2O \rightarrow 3CO_2 + 20H^+ + 20e$$
Propane

A gaseous fuel of a rather different type is ammonia, which is oxidized to nitrogen:

$$2NH_3 + 6OH^- \rightarrow N_2 + 6H_2O + 6e$$

Liquid fuels offer some advantages, e.g. the electrode system is usually simpler and storage is less of a problem. Apart from higher molecular weight hydrocarbons, methyl alcohol has been used:

$$CH_3OH + H_2O \rightarrow CO_2 + 6H^+ + 6e$$
Methyl alcohol

It can be seen that a variety of materials can be used as fuels for these cells, but although they have some attractive features, e.g. high efficiency and no need of recharging, a number of technical difficulties must be overcome before their use is widespread.

8.6. Corrosion

A knowledge of the causes of corrosion is of great importance to anyone involved in the engineering sciences. Corrosion is defined here as the transformation of a metal from the elemental to the combined state, although in its broadest sense the word could also be applied to the deterioration of any material. Under the appropriate conditions all metals will undergo chemical attack leading ultimately to severe deterioration of strength or of appearance.

The study of metallic corrosion is complicated by two principal factors. Firstly, corrosion is rarely a simple process and the interplay of a variety of possible mechanisms must be considered. Secondly, corrosion is, in general, a fairly slow process and accelerated tests have to be devised. This leads to the difficulty of deciding to what extent conclusions drawn from such tests are valid for the slow naturally occurring reaction.

Although it is probably true that all types of corrosion could be described as electrochemical reactions, it is convenient to classify corrosion into two main types, *direct* and *indirect*. In the former case we consider a localized reaction to occur, whereas in the latter, galvanic cells are established such that the complementary oxidation and reduction reactions can take place in different parts of the system.

8.7. Direct corrosion

Direct chemical attack of a metal is usually a fairly straightforward reaction and relatively easy to control. In this type of attack all chemical reaction must occur at the surface of the metal, and the corrosion product is deposited there. Since the reaction is a surface one, corrosion can continue only if the attacking species can successfully diffuse through the layer of reaction product. Because of this many direct corrosive reactions are self-inhibiting.

The most important type of direct attack is *atmospheric corrosion*. The main constituents in the atmosphere which can cause corrosion are oxygen, water and carbon dioxide, although in industrial atmospheres the presence of soot and sulphur compounds can cause severe attack. Direct oxidation of most metals occurs only to a slight extent at normal temperatures, but as the temperature is increased they will oxidize readily. In this respect we can classify metals into two main categories, those which form oxide films with a volume less than that of the underlying metal, and those which form an oxide film of greater volume.

The metals of the first two groups of the periodic table are of the first type, where the corrosion product has an open porous structure and is therefore not a protective film. These metals then are very susceptible to aerial oxidation. Metals of the second type may oxidize readily at first, but since the corrosion product hinders the diffusion of oxygen inwards (or the diffusion of metal ions outwards) the rate of corrosion often decreases markedly.

A complicating factor must now be considered. The protective oxide layer is effective because it occupies a greater volume than the underlying metal. It is quite likely therefore that this protective layer is under considerable lateral stress, and cracking of the oxide film may occur, particularly when the film is thick. If this happens the metal is no longer protected and corrosion may then proceed once again. These types of corrosion are illustrated in Fig. 8.5, where curve *A* shows how corrosion proceeds when non-protective films are produced, while curve *B* is typical of protective film formation followed by rupture of the protective coating.

As mentioned already, humidity is an important factor in determining the extent of corrosion. In general, the rate of

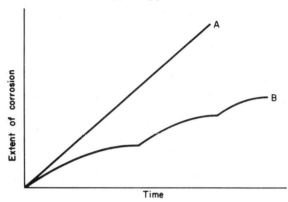

Fig. 8.5. Rate of atmospheric corrosion.

corrosion increases with increasing humidity. Indeed, for many metals a *critical humidity* exists, below which corrosion occurs relatively slowly, but above which a marked acceleration of corrosion rate occurs.

We must also remember that in industrial atmospheres the presence of soot and sulphur compounds is important. Sulphur compounds are partially oxidized by the air to form highly corrosive sulphuric acid. The corrosive effect is even more pronounced if soot particles are present, because moisture and sulphur dioxide are readily absorbed by carbon, thus forming locally very high concentrations of corrosive materials.

8.8. Indirect corrosion — simple galvanic cells

The process of corrosion is said to be indirect if it can be divided into two separate reactions, an oxidation and a reduction, both involving the transfer of electrons. We learned earlier in this chapter that this is just what occurs in a simple cell. Obviously such a situation may often arise, i.e. different metals or alloys will be in contact and if immersed in a suitable electrolyte will function as a galvanic cell. The more active metal will act as the anode and will be oxidized, while the more noble metal will act as a cathode. Thus at the anode we expect reactions of the type

$$M \rightarrow M^{2+} + 2e$$

whilst at the cathode hydrogen gas may be evolved, particularly from acidic solutions

$$2H^+ + 2e \rightarrow H_2$$

While the current continues to flow corrosion will persist, but the hydrogen gas may accumulate at the cathode and cause *polarization*, thus preventing further corrosion. In such cases corrosion can only continue if the cell is depolarized by chemical removal of the hydrogen, e.g. by the action of dissolved oxygen or some other oxidizing agent. The rate of reaction is then determined by the rate at which oxygen diffuses to the cathode, and hence corrosion is in these circumstances fairly slow.

This type of corrosion is extremely common where metals are riveted or screwed together. Since the severity of corrosion depends on the current density, a large cathode area coupled to a small anode area will lead to severe corrosion. In a conducting medium a steel rivet in a copper sheet will be quickly corroded because of this effect. Corrosion of this type can usually be prevented by suitably insulating the junction of the two metals and preventing the current from flowing.

A different example of the same phenomenon is found in the effect of impurities in metal. Very pure zinc dissolves extremely slowly in non-oxidizing acids because the over-voltage effect prevents the easy evolution of hydrogen gas. If iron is present as an impurity, however, a series of galvanic cells is produced such that the hydrogen gas is readily evolved from the cathodic impurities and the zinc, acting as the anode, then readily dissolves.

In aqueous solution there is always the possibility that the anode reaction may involve the discharge of hydroxyl ions rather than the dissolution of metal:

$$M \rightarrow M^{2+} + 2e$$
$$OH^- \rightarrow OH + e$$

If this occurs, a protective layer of oxide or hydroxide may be deposited on the metal surface, preventing further corrosion. The metal is said to be rendered *passive*.

We must remember therefore that oxygen can act in two possible ways in corrosion. It may accelerate corrosion by depolarizing the cathode, or it may inhibit the anodic reaction by rendering the metal passive.

8.9. Concentration cells

We learned in Section 8.3 that when metals are in contact with electrolytic solutions of differing concentrations galvanic action will result if the metals are connected by a suitable conductor. The metal in contact with the more dilute solution will dissolve, the cell current being such that it decreases the cell potential. Metal ion concentration cells of this type are quite common since a concentration gradient can arise in a variety of

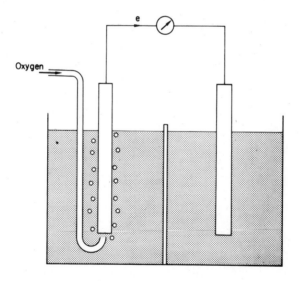

FIG. 8.6. Differential aeration cell.

ways. In particular, recessed stagnant areas are very often places where the metal ion concentration can accumulate.

Probably the most commonly encountered concentration cell is the oxygen concentration cell, usually termed the *differential aeration cell*. We know that oxygen can act as a depolarizing agent for the cathode of a galvanic cell, and can also produce protective oxide coatings. Consequently a metal surrounded by a high oxygen concentration tends to become more 'noble', or cathodic, and conversely areas with low oxygen concentration become anodic.

This type of concentration cell can be readily demonstrated by assembling two identical metal electrodes as illustrated in Fig. 8.6, such that one electrode compartment can be aerated. When air is bubbled over one electrode surface a galvanic cell is established and a current flows, the aerated rod being cathodic, the unaerated anodic. The cell can be reversed by bubbling air around the other electrode. At the cathode hydroxyl ions are produced, while at the anode the metal dissolves.

$$\text{Cathode} \qquad\qquad \text{Anode}$$
$$\tfrac{1}{2}O_2 + H_2O + 2e \rightarrow 2OH^- \qquad M \rightarrow M^{2+} + 2e$$

This type of concentration cell can lead to corrosion in a variety of situations. If, for example, we imagine a sheet of iron which has developed a small crack, then the bulk of the metal surface is more efficiently aerated than the small crevice and consequently tends to be cathodic. Galvanic corrosion will occur, and as this proceeds iron hydroxide ($Fe(OH)_2$) will be formed as a product of the ferrous ions diffusing from the anode and the hydroxyl ions formed in the cathode reaction. This solid corrosion product tends to be deposited in such a way that it enhances the differential aeration effect, and corrosion becomes more severe. It is this tendency for the corrosion product to be formed at a distance from the point of active corrosion that can make galvanic corrosion so much more serious than direct chemical attack, because there is little possibility of forming a protective layer.

Another example of this type of corrosion is waterline attack where severe corrosion can occur at a liquid–air interface. The oxygen concentration in the bulk of the liquid is much less than that of the surface layer and again a corrosion cell is established.

8.10. Stress corrosion

We can consider stresses to be of two main types, a static stress or a repeated stress. When the former type is coupled with corrosion we find the phenomenon of *stress corrosion*, while the latter type leads to *corrosion fatigue*.

While it is true that stressed metal is rendered less noble than unstressed and thus galvanic action is possible, the effect is only slight and serious corrosion does not normally occur. If, however, some degree of cracking occurs the coexistence of corrosive action and stress can lead to rapid deterioration of the metal. Sometimes the failure is mainly mechanical, the function of corrosion being simply to initiate crack formation or to destroy the sound material which limits the amount of cracking. On other occasions a steady electrochemical action occurs, the stress preventing the formation of any protective film and concentrating the point of attack at the very tip of the crack.

8.11. Protection against corrosion

The corrosion of metal structures can be prevented, or at least slowed down, by a number of techniques, one of which is the use of *inhibitors*. It has already been pointed out that one of the characteristics of galvanic corrosion is that the anodic and cathodic reactions can occur at different parts of the system and that therefore no protective film of corrosion product is formed. The oxidation and reduction are, however, mutually dependent and if one of these reactions can be prevented then the overall corrosion reaction must cease. Corrosion inhibitors therefore are chemicals which act either at the anode or the cathode to hinder the corrosive reaction.

Chemicals which function at the anode are termed *anodic inhibitors*, and can be either organic or inorganic reagents. A large number of inorganic compounds have been found to act as inhibitors, but the most common are sodium carbonate, sodium hydroxide, sodium phosphate, sodium silicate and potassium chromate. The mechanism whereby these inhibitors render the anode passive is not altogether certain, but all seem to be capable of forming insoluble films on the metal surface. The effective film may not be an insoluble phosphate or carbonate, but may be a hydroxide layer since most of the common inorganic inhibitors yield alkaline solutions.

Organic materials can also act as anodic inhibitors, e.g. thiophenol $(C_6H_5 \cdot SH)$ attaches itself through the electronegative sulphur atom thus rendering that particular metallic region more negative and so unlikely to corrode. The degree of protection can be enhanced by introducing suitable substituents in the benzene ring.

The use of anodic inhibitors can be a dangerous technique. The addition of inhibitor simply restricts the area at which corrosion occurs, and it is therefore possible if the correct amount of inhibitor is not added that very intense attack could occur at the small anodic areas which were not protected. Apart from this danger, many of the inorganic inhibitors are oxidizing agents and can therefore act as depolarizers and consequently increase the rate of corrosion.

Some inorganic salts act as *cathodic inhibitors*, e.g. zinc, magnesium and calcium salts. This effect is generally due to the formation of relatively thick insoluble films which hinder the diffusion of ions at the cathode and thus slow the corrosive reaction down.

Organic inhibitors can also be used to form protective layers at the cathode by adsorption. Organic bases, e.g. pyridine or other amines, form positive ions which attach themselves through the amino nitrogen atom to the cathode areas, forming protective films. The efficiency of this type of inhibitor depends on its 'covering power', e.g. high molecular weight amines are more effective than those of low molecular weight, and highly branched tertiary amines are better inhibitors than straight chain primary ones.

A second way of protecting against corrosion is by the use of protective coatings, which may be *metallic*, *inorganic* or *organic*.

Metallic coatings are commonly used and can be applied in many different ways. At first sight the most obvious choice for the coating material would be a noble metal, because of their known chemical resistance. While these are used for some special purposes, e.g. silverplating, they suffer two main disadvantages, viz. their high cost and their tendency to set up extremely corrosive galvanic cells at any discontinuity in the protective coating.

Commonly used metal coatings are chromium, tin and zinc. Chromium plating—a thin coating of chromium electrolytically deposited over a layer of nickel—is widely used because of its hardness, attractive lustre and tarnish resistance. Chromium is itself highly resistant to corrosion, readily becoming passive due to protective layer formation, but because of this it permits the formation of galvanic cells with underlying less noble metals, e.g. steel. It is important therefore that a strongly adhering film free of cracks be deposited on the metal to be protected.

Zinc and tin are rather similar in that both can be deposited either electrolytically or by a dipping process. Both metal coatings are resistant to general corrosion and are anodic to iron.

Examples of inorganic and organic protective coatings are found in *enamels* and *paints*. Here the obvious mechanism is simply to protect the underlying metal by a coating of some stable material which is impermeable to moisture. In fact the process is not so simple as this, and it is essential to pretreat the metal with an inhibitor before applying the outer protective coat, e.g. 'priming' with inhibitive pigments such as red lead or zinc chromate.

The final method of protection to be discussed is termed *cathodic protection*, where inhibition is achieved by using the current produced by galvanic action to protect vital parts of the system. This is illustrated in Fig. 8.7. Initially let us imagine that the cell consists simply of two electrodes A and C, the latter being the more noble. If galvanic action were to occur, metal ions would pass into solution at the anode, and electrons would flow from the anode to the cathode as shown.

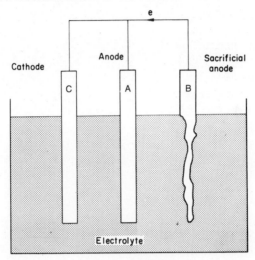

FIG. 8.7. Sacrificial anode.

If we now add a third electrode B to the system, where this electrode B is a metal even more reactive than A, a corrosion cell still functions but the electrons now flow from the anode B. All corrosion is confined to the electrode B which is described as a *sacrificial anode*. This is designed to be easily renewed when necessary, e.g. magnesium bars bolted to the sides of ships act as sacrificial anodes and thus protect the hulls from corrosion.

It can be seen from Fig. 8.7 that the function of the anode B is to maintain a supply of electrons to the cathode. This protective current could be supplied in another way, viz. from a d.c. generator supplying current to a non-sacrificial anode. A current supplied in this way is termed an *impressed current*.

Both of these methods are widely applied to control corrosion of large metal structures such as ships, pipelines, etc., and are normally used to supplement the normal types of protective coatings described previously.

CHAPTER 9

The Metallic Elements

9.1. Extraction of metals

In a text of this size it is not possible to consider in any detail the chemistry of all the elements in the periodic table. Instead, in this chapter, we shall study a selection of elements, and in particular the more important metals. This being the case, we shall start by considering the occurrence of metals, and briefly discuss how they can be obtained in a pure state.

We have already learned (Chapter 8) of the significance of the electrochemical series and therefore can appreciate that only a very few metals can be expected to exist naturally in a pure state. Those that do, viz. metals below hydrogen in the series, such as copper, silver, gold and platinum, are termed *noble metals*.

All other metals are found chemically combined with other elements, the natural materials from which the metals can be economically extracted being termed *ores*. The process of obtaining a pure metal from an ore generally involves three distinct processes:

(i) *preliminary treatment*,
(ii) *reduction*,
(iii) *purification*.

The preliminary treatment of an ore does not generally involve any chemical properties of the material, but rather depends on the physical characteristics of the ore. Typically, the initial treatment would involve pulverizing the ore followed by removal of waste material. If the ore particles are fairly dense, the concentration of the ore may be achieved simply by washing, thus removing the less dense waste.

The converse process of *flotation* is sometimes used, particularly with sulphide ores. In froth flotation a froth is formed by blowing air into a pulp of pulverised ore and water to which a small amount of oil has been added as a frothing agent. Minerals that have an affinity for the air bubbles rise to the surface in the froth, while the waste material (e.g. sand, clay) is more readily wetted by the water and sinks to the bottom. The process has been considerably refined by the development of reagents specific to one mineral or class of minerals which form water-repellent films on the mineral surface. These reagents are usually polar organic molecules (e.g. amines, phenols, xanthates) which attach themselves with their polar ends to the mineral surface thus producing a non-polar surface film around the ore particle, which is then not readily wetted by water. In this way it is even possible to separate one ore from another by the correct choice of suitable reagents.

Before the second stage, some ores undergo a *roasting process* to convert them to a compound more suitable for the reduction step, e.g. carbonates and sulphides are converted to the corresponding oxides by heating.

$$CaCO_3 \rightarrow CaO + CO_2$$
$$2ZnS + 3O_2 \rightarrow 2ZnO + 2SO_2$$

The reduction of the ore to the crude metal can be carried out in a variety of ways, depending on the chemical properties of the metal and the general economics of the process. Generally, oxide ores of metals which are not too firmly bound are reduced with carbon, or in some cases with reducing agents such as aluminium or hydrogen:

$$ZnO + C \rightarrow Zn + CO$$
$$Cr_2O_3 + 2Al \rightarrow 2Cr + Al_2O_3$$

The ores of the active metals which are high in the electrochemical series, e.g. the metals of groups I and II in the periodic table, are not so easily reduced by chemical means. In these cases an electrolytic technique is usually adopted.

Purification, or *refining*, can also be achieved in a variety of ways. For example, if the metal has a low boiling point, purification may involve only a simple distillation, although the most common method is probably electro-refining where the impure metal is used as the anode of a suitable electrolytic cell. Metals of high purity can be obtained by *zone refining*. Here a narrow molten zone of metal is made to move slowly along the length of the material as shown in Fig. 9.1, the impurities being swept

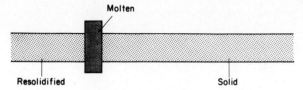

FIG. 9.1. Zone refining.

to one end of the rod. A number of sweeps may be necessary to obtain the desired purity. This method is based on the fact that when a liquid mixture is cooled, the solid separating out usually has a different composition from the liquid. Since the impurities are generally preferentially dissolved in the liquid zone, the re-crystallized solid will contain less of the impurity. Other classical methods of separation such as *ion exchange* or *solvent extraction* may be used. Occasionally the crude metal is reacted to form a compound which is more easily purified than the metal itself. Nickel, for example, is readily converted to a volatile carbonyl compound:

$$Ni + 4CO \rightarrow Ni(CO)_4$$

The nickel carbonyl, which boils at 43°C, can be distilled off and decomposed at higher temperatures (150–200°C) to yield a high purity metal.

9.2. Group IA elements

The elements of group IA in the periodic table are lithium, sodium, potassium, rubidium, caesium and francium, and are commonly referred to as the *alkali metals*. Their electronic configurations and some of their physical properties are shown in Table 9.1.

TABLE 9.1. THE ALKALI METALS

Element	Electronic configuration	Atomic radius (Å)	M.p. (°C)	B.p. (°C)
Li	2.1	1·55	186	1336
Na	2.8.1	1·90	97·6	880
K	2.8.8.1	2·35	62·3	760
Rb	2.8.18.8.1	2·48	38·5	700
Cs	2.8.18.18.8.1	2·67	28·5	670

The alkali metals are the most electropositive elements known and very readily lose one electron to form a monopositive ion:

$$M \rightarrow M^+ + 1e$$

This is so for two reasons. From Table 9.1 it can be seen that the electronic configurations of the alkali metals are such that each metal has one electron in excess of an inert gas configuration, therefore by losing a single electron a very stable configuration is attained. The size of the atoms is also an important factor. The alkali metals have a very large atomic radius and consequently the outermost electrons are not firmly held by the nucleus and are fairly easily lost. Because of this ease of loss of the valency electron the alkali metals are extremely reactive and will react with even the weakest oxidizing agents. This reactivity is reflected in the high position these metals occupy in the electrochemical series.

The reactivity of these metals causes problems of various kinds; for example, there is no possibility of finding these metals in the uncombined state in nature. On the contrary, they are generally found as complex silicates, e.g. $LiAlSi_2O_6$, $KAlSi_3O_8$, or at least as very stable salts which are difficult to reduce. Because of their high chemical reactivity the general method of obtaining the metal is by electrolysis of a suitable salt. Sodium, for example, is usually manufactured by the *Downs process*, where a mixture of fused sodium chloride and calcium chloride is electrolysed using a carbon anode and a copper or iron cathode. Since sodium chloride has rather a high melting point (801°C) the calcium chloride is added to reduce the melting point of the electrolyte to around 600°C. Sodium is produced at the cathode:

$$Na^+ + 1e \rightarrow Na$$

while chlorine gas is evolved at the anode:

$$2Cl^- \rightarrow Cl_2 + 2e$$

Again at this stage of the extraction process problems arise because of the extreme reactivity of these metals. The electrolysis cell must be so designed that the metal produced does not immediately react to form a compound either with the by-product of the electrolytic reaction or with air. Once obtained, the metals are usually stored under kerosine.

The alkali metals crystallize with a body-centred cubic lattice (see Chapter 6) and are all extremely soft, malleable and ductile. They are good conductors of both heat and electricity. The most important members of the group are sodium and potassium, but from what has been said above it will be realized that they cannot readily be used as conventional metals because of their high chemical activity. The pure metals tarnish very quickly in air due to the formation of oxide films, and react violently with water with the evolution of hydrogen gas;

$$2M + 2H_2O \rightarrow 2M^+ + 2OH^- + H_2$$

The metals are used technically for some rather specialized purposes, e.g. liquid sodium and potassium alloys have been used as heat transfer agents in nuclear reactors. Sodium in particular is widely used in metallurgy and forms alloys with many metals. Sodium dissolves in mercury to form an *amalgam* which is a commonly used reducing agent in chemical reactions. Its main advantage is that the presence of the mercury makes the sodium far less reactive so that the reactions are more easily controlled.

Of the remaining elements in the group, lithium is the most common. Again its main use is in metallurgy where it is used in metal refining because of its ability to combine with oxygen and nitrogen, and in the formation of alloys with metals such as magnesium and aluminium.

Caesium and rubidium are not very abundant in nature and have a chemical reactivity even greater than that of sodium or potassium. They have, however, a very useful property in that they can be induced to ionize, i.e. to eject an electron, simply by stimulation with light. This is termed the *photoelectric effect*, and these metals are used in the manufacture of *photocells*, which are devices which utilize this property to convert a light signal into an electric pulse.

The last member of the series, francium, occurs to only a minute extent as a short-lived radioactive isotope.

Technically the most important alkali metal compounds are the hydroxides and carbonates, in particular sodium hydroxide (*caustic soda*) and sodium carbonate, commonly referred to as *soda ash*, or simply *soda*. Both these chemicals are extensively used in the manufacture of other chemicals, textiles, soap, paper and in petroleum refining. Their manufacture will be fully discussed in Chapter 14.

9.3. Group IIA elements

The group IIA elements, beryllium, magnesium, calcium, strontium, barium and radium, which are often referred to as the *alkaline earth* elements, differ from the group IA metals in that they all possess two valency electrons, as can be seen from Table 9.2.

We have already learned that atoms which are fairly large and which possess a small number of electrons in excess of a stable electronic configuration generally lose their electrons readily to form ionic compounds. This is the case with the

TABLE 9.2. THE ALKALINE EARTH METALS

Element	Electronic configuration	Atomic radius (Å)	M.p. (°C)	B.p. (°C)
Be	2.2	0·89	1280	2970
Mg	2.8.2	1·36	650	1107
Ca	2.8.8.2	1·74	842	1240
Sr	2.8.18.8.2	1·91	800	1150
Ba	2.8.18.18.8.2	1·98	850	1140
Ra	2.8.18.32.18.8.2	2·00	700	1140

alkaline earth metals which are similar in behaviour to the group IA elements, except that they are not quite so reactive. The early members of the group, beryllium and to some extent magnesium, are much smaller than the later members and behave rather differently in that they tend to form covalent bonds.

From their position in the electrochemical series it can be seen that the alkaline earth metals will be found only in the combined state in nature, in compounds which will be fairly difficult to reduce by chemical means. Like the alkali metals, these metals are found as complex silicates (e.g. $Mg_3Si_4O_{10}(OH)_2$), but also as insoluble carbonates and sulphates. As with the group IA metals, reduction of the ore is generally achieved by an electrolytic process. In the extraction of magnesium from sea water, for example, the magnesium is first precipitated as the hydroxide, converted to the chloride by treatment with hydrochloric acid and the dry chloride, fused with other salts to lower the melting point, electrolysed at about 700°C to yield high purity metal.

Beryllium and magnesium

Of these two metals magnesium is the more important technically. Although beryllium is used in some specialized alloys, its compounds are highly toxic and thus not suitable for general use.

Magnesium is the lightest of all structural metals, but since it is such a reactive metal its corrosion resistance is not very high. It has, however, a good resistance to atmospheric corrosion because of the formation of protective films. Since it is a very light metal and is easily cast and welded it is widely used as a constructional material, particularly in the form of alloys, these with aluminium and zinc being the most important.

Chemically, a useful feature of magnesium is its great affinity for oxygen. Because of this magnesium can be used to reduce metal oxides to the corresponding metal.

Calcium, strontium and barium

These metals are more reactive than beryllium or magnesium, and are readily attacked both by oxygen and water, and therefore find no application as structural materials. Of the three, calcium is the most widely used, mainly in metallurgy where it is used as an alloying agent. Because of its high chemical reactivity, e.g. the ease of formation of oxides, nitrides, etc., calcium is also used to remove impurities from metals.

The most important compounds of the group IIA metals are the oxides, hydroxides and sulphates. Calcium oxide (quicklime) is obtained by thermal decomposition of the carbonate. The oxide formed is readily converted to the hydroxide (slaked lime) by treatment with water.

$$CaCO_3 \rightarrow CaO + CO_2$$
$$CaO + H_2O \rightarrow Ca(OH)_2$$

Since it is considerably cheaper than sodium hydroxide, lime is widely used as an industrial base and is an important constituent of cements (Chapter 7).

Calcium sulphate occurs in nature as *anhydrite* ($CaSO_4$) and as *gypsum* ($CaSO_4 \cdot 2H_2O$). Gypsum is widely used, for example, in the manufacture of cements, paper, paints and pharmaceuticals. On heating, gypsum loses part of its water of crystallization to form *plaster of Paris*:

$$CaSO_4 \cdot 2H_2O \rightleftharpoons CaSO_4 \cdot \tfrac{1}{2}H_2O + \tfrac{3}{2}H_2O$$

The use of plaster of Paris for making casts and moulds arises from the reversibility of this reaction. Since the reverse reaction is accompanied by an increase in volume, very good reproductions of the mould are obtained.

9.4. Group IIIA elements

This group of the periodic table is composed of the elements boron, aluminium, gallium, indium and thallium, whose physical properties are listed in Table 9.3.

TABLE 9.3. THE GROUP IIIA METALS

Element	Electronic configuration	Atomic radius (Å)	M.p. (°C)	B.p. (°C)
B	2.3	0·80	2300	2550
Al	2.8.3	1·25	660	2300
Ga	2.8.18.3	1·25	29·8	2000
In	2.8.18.18.3	1·50	155	1450
Tl	2.8.18.32.18.3	1·55	304	1460

Since it becomes progressively more difficult for an element to transfer electrons, these elements differ from groups IA and IIA in that their compounds tend to be covalent rather than ionic in character, and their general properties differ in many respects from those of the preceding groups. Boron in particular differs from the remainder of the group in that it is a non-metal. This change in character as we descend the group is illustrated by the behaviour of the hydroxides of the elements. The hydroxide of boron $(B(OH)_3)$ is acidic, the hydroxides of indium and thallium are basic, while the hydroxides of aluminium and gallium are *amphoteric*, i.e. they can act either as a base or as an acid, depending on the reaction conditions.

Aluminium, which is by far the most important member of the group, occurs widely in the form of complex silicates (e.g.

$KAlSi_3O_8$) from which it would be difficult to extract the metal, and as *bauxite* ($Al_2O_3 \cdot xH_2O$) which is the important ore. In the extraction process use is made of the amphoteric nature of aluminium. The concentration of the ore is achieved by treatment with caustic soda solution where the aluminium dissolves with the formation of the *aluminate* ion,

$$Al_2O_3 \cdot xH_2O + OH^- \rightarrow [Al(OH)_4]^-$$

which is subsequently precipitated as the hydroxide ($Al(OH)_3$). This is in turn converted to the oxide from which the aluminium may be obtained by electrolysis, using an electrolyte of aluminium oxide dissolved in molten fluorides.

Aluminium metal is a very good conductor of both heat and electricity. Although itself a fairly reactive element, its corrosion resistance is high because of the formation of protective oxide coatings. This property, together with its lightness, makes the metal useful as a structural material. The pure metal is fairly soft, but it very readily forms alloys with, for example, copper, manganese or magnesium to give materials of good mechanical properties. The metal is also a very efficient reflector of heat and light and is therefore widely used as an insulating material.

Like magnesium in the previous group, aluminium has a very high affinity for oxygen, and this property can be utilized to obtain certain metals from their oxides when common reducing agents such as carbon or hydrogen cannot be used, e.g. the reduction of the oxides of manganese, chromium and tungsten. An extension of this is the *thermite reaction* where a mixture of powdered aluminium and ferric oxide is ignited:

$$2Al + Fe_2O_3 \rightarrow 2Fe + Al_2O_3$$

The large amount of heat evolved in this reaction is sufficient to melt the mixture, so that this process can be used for welding operations.

The remaining metals in the group are fairly rare, and are generally obtained as by-products from the manufacture of aluminium or zinc. Perhaps the most important common property of these metals is that they form compounds with arsenic and antimony to give semiconducting materials, of great importance to the electronics industry.

9.5. Group IVA elements

The group IVA elements, carbon, silicon, germanium, tin and lead show an even more pronounced change from nonmetallic to metallic properties than do the group IIIA elements. Carbon and silicon are predominantly nonmetallic, tin and lead are typical metals and germanium is somewhat intermediate in character.

TABLE 9.4. THE GROUP IVA ELEMENTS

Element	Electronic configuration	Atomic radius (Å)	M.p. (°C)	B.p. (°C)
C	2.4	0·77	3500	4200
Si	2.8.4	1·17	1420	2355
Ge	2.8.18.4	1·22	959	2700
Sn	2.8.18.18.4	1·40	232	2270
Pb	2.8.18.32.18.4	1·75	327	1620

The first three members of the group tend to form typical covalent bonds, whereas tin and lead form normal cations. The pronounced change in character as we descend the group is reflected in the melting points. Carbon, silicon and germanium form strong interlocked covalent structures, whereas tin and lead are typical low melting metals. From their position in the electrochemical series we can see that compounds of tin and lead should be easily reduced by chemical means. This, together with the fact that they are low melting, has meant that these metals were amongst the first to have been used by man.

Tin

The principal ore is *cassiterite* (SnO_2), which is first roasted to remove arsenic and sulphur, and finally reduced by carbon in a blast furnace:

$$SnO_2 + 2C \rightarrow Sn + 2CO$$

Refining of the metal is generally achieved by electrolysis.

Tin exhibits the phenomenon of *allotropy*, i.e. it can exist in more than one form:

$$\text{Grey tin} \underset{}{\overset{13.2°C}{\rightleftharpoons}} \text{White tin} \underset{}{\overset{161°C}{\rightleftharpoons}} \text{Brittle tin}$$

The generally used form of tin is white tin, but we see that this converts to grey tin, which is a powdery form of the element, at temperatures below 13·2°C. In practice this change occurs very slowly, but articles made of tin can be expected to disintegrate if exposed to low temperatures for prolonged periods. Nevertheless, tin is a very versatile metal with some useful properties. Although a fairly weak metal it is easily worked, has a high resistance to corrosion, is non-toxic and has a pleasing appearance. For these reasons the metal is widely used as a surface coating, as was described in Chapter 8.

The other main application for tin is its use in making alloys with a high resistance to corrosion. Three main types exist:

   (i) *bronzes* containing copper and tin;
  (ii) *solders* containing lead and tin;
 (iii) *white metals* containing lead, antimony, copper and tin. These are used for bearings because of their strength and good antifriction properties.

The nature of the material depends on the composition of the alloy. So-called *tinman's solder* (35 per cent Pb) is solid at 183°C, but liquid at 184°C, and is used, for example, in soldering electrical equipment where a high fluidity is desirable. *Plumber's solder*, on the other hand, does not melt sharply at any given temperature, but rather becomes plastic over a temperature range (183–250°C) and is therefore easily shaped and manipulated.

Lead

The principal lead ore is *galena* (PbS), which is first roasted to form the oxide and then reduced by carbon. The metal is soft, dense, low melting and very easily worked. Like tin its resistance to corrosion is very high due to the formation of protective carbonate coatings. For these reasons lead has been widely used as a cable sheathing, for roofing, piping, battery grids and as a corrosion lining in chemical plant. Because of its high density, it is also used as a shielding material against radiation. Since lead and its salts are highly toxic, it is no longer used as a piping material for drinking water. Soft waters generally attack lead sufficiently to make its use for this purpose dangerous.

Like tin, lead is widely used in alloys such as solders, bearing metals and antimony alloys, where the incorporation of a small amount of antimony increases the hardness and strength of the lead. Both tin and lead show variable valency, forming both two- and four-valent compounds. Thus lead dioxide (PbO_2), which is the main constituent of the anode of the lead storage battery, is reduced during the discharge reaction to plumbous ions (Pb^{2+}), as explained in Section 8.4.

Another oxide of lead which is of technical importance is *red lead* (Pb_3O_4), which is widely used as an undercoat for painting steel structures. Since red lead is an oxidizing agent it renders the iron passive and so helps to prevent corrosion. Other lead compounds which are used as pigments are *white lead* (Pb_3-$(OH)_2CO_3$) and *lead chromate* ($PbCrO_4$). Since lead salts are highly toxic, however, their use is being superseded by other pigments.

The last member of the group which can be classed as a metal is germanium. Chemically this is a very unreactive element and is used mainly by the electronics industry in the manufacture of transistors.

9.6. Early transition elements — groups IIIB–VIIB

We have already learned that the transition metals are those which possess incomplete inner electron shells, and that the first transition series starts at the element scandium and finishes with copper. The element zinc, however, has much in common with the transition metals and will therefore be considered under this general heading. Since the metals of the first transition series are by far the most abundant we shall consider their properties in some detail and refer only briefly to a few of the heavier transition metals.

As can be seen from Table 9.5 the elements we are considering here are scandium, titanium, vanadium, chromium and manganese. It will be noted that with the exception of chromium each metal contains two outer electrons, and we would expect therefore that the chemical behaviour of these elements would be rather similar. In the case of chromium greater stability is attained by accommodating one of the $4s^2$ electrons in a $3d$ orbital so that each of the $3d$ orbitals contains one electron.

TABLE 9.5. EARLY TRANSITION METALS

Element	Electronic configuration	Atomic radius (Å)	M.p. (°C)	B.p. (°C)
Sc	2.8.8.1.2	1·44	1200	2400
Ti	2.8.8.2.2	1·32	1800	3000
V	2.8.8.3.2	1·22	1710	3000
Cr	2.8.8.5.1	1·17	1890	2480
Mn	2.8.8.5.2	1·17	1260	1900

The first transition metal of real technical importance is titanium, which is a fairly abundant element found in the ores *ilmenite* ($FeTiO_3$) and *rutile* (TiO_2). The metal itself is difficult to obtain in a state of high purity because of its high melting point and because it reacts readily with non-metals such as oxygen, nitrogen, hydrogen and carbon. It can, however, be obtained by the reduction of titanium halides with active metals such as sodium or magnesium. As a metal it is strong, light and has excellent corrosion resistance, and is consequently widely used in making alloys suitable for the construction of jet engines and aircraft. The most important compound is the dioxide (TiO_2) which is an excellent white pigment and is now very widely used in the manufacture of paints.

The other members of the titanium subgroup are zirconium and hafnium, which like titanium have a high resistance to corrosion. Both these metals are used in nuclear engineering, zirconium because of its low neutron capture cross-section being used as a sheathing material for fuel elements, and hafnium because of its capacity for absorbing neutrons, as control rods in a nuclear reactor.

The vanadium subgroup, which includes the elements niobium and tantalum, are of minor importance, their main use being in the formation of alloys with iron to yield steels of special properties, and need not be discussed further here.

Chromium occurs mainly as the ore *chromite* ($FeCr_2O_4$) from which it is normally extracted as an alloy with iron (ferrochromium) simply by reduction with carbon. The pure metal can be obtained by reduction of the oxide with aluminium powder.

$$FeCr_2O_4 + 4C \rightarrow Fe + 2Cr + 4CO$$
$$Cr_2O_3 + 2Al \rightarrow Al_2O_3 + 2Cr$$

The main uses of chromium are in the formation of electroplated coatings or, as with so many of the early transition metals, in the formation of high strength corrosion resistant alloys. Chromium itself is a very hard metal with a high metallic lustre which owes its resistance to corrosion to the formation of a stable protective oxide film. The metal alloys readily with iron and nickel to form stainless steels, and with nickel to form an alloy with excellent high temperature characteristics (nichrome), which is used in electrical heating devices.

The other members of the chromium subgroup are molybdenum and tungsten, the former being an important component of steels which retain their strength at high temperatures. Tungsten is one of the hardest and heaviest of metals, is the highest

melting metal, and chemically is highly inert. As an alloy with iron it is used in the production of steels with good temperature characteristics, and as the pure metal is used as filaments in electric lamps.

Manganese, the final member of this series, occurs in the ore *pyrolusite* (MnO_2) from which the pure metal may be obtained by reduction with aluminium.

$$3MnO_2 + 4Al \rightarrow 3Mn + 2Al_2O_3$$

Again, this metal is widely used in the manufacture of steels where it is used as a 'scavenger' by combining with oxygen and sulphur, and also to form an alloy with iron which is both hard and tough.

Perhaps the most important compound is manganese dioxide which is used as a decolourizing agent in the glass industry, as a depolarizer in dry cells, and for the production of other manganese salts.

9.7.　Group VIIIB elements

The so-called group VIIIB metals are, strictly, three separate groups which are considered together because of their similarity. Thus the metals iron, cobalt and nickel, whose physical properties are listed in Table 9.6, are often described as the *first transition triad*.

TABLE 9.6. GROUP VIIIB METALS

Element	Electronic configuration	Atomic radius (Å)	M.p. (°C)	B.p. (°C)
Fe	2.8.8.6.2	1·16	1535	3000
Co	2.8.8.7.2	1·16	1495	2900
Ni	2.8.8.8.2	1·15	1455	2900

Apart from similarities of chemical and physical properties, these elements are unique in that they are *ferromagnetic*, i.e. they are strongly attracted to magnetic fields and can themselves be permanently magnetized.

Iron is the most important of all the metals and is by far the most widely used. It is the second most abundant metal in the earth's crust and can be obtained from its ore by a simple reduction treatment with carbon. The metal itself is very versatile with good mechanical properties, which can be further improved by the addition of small amounts of 'impurities' to form alloys. Unfortunately the metal is fairly reactive chemically and hence is particularly susceptible to corrosion.

The most important source minerals are *hematite* (Fe_2O_3) and *magnetite* (Fe_3O_4), which are reduced in a blast furnace (Fig. 9.2). The charge of ore, carbon coke and flux (limestone) is introduced at the top of the furnace, where the temperature is around 200°C. Hot air (in modern practice this is enriched

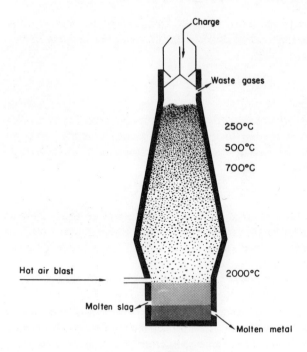

FIG. 9.2. Blast furnace.

with oxygen and fuel oil) is forced into the base of the furnace where the coke reacts to form carbon monoxide, at the same time liberating a large quantity of heat.

$$2C + O_2 \rightarrow 2CO$$

This region of the furnace has the highest temperature (2000°C) and as the gases rise up the furnace they eventually cool to around 200°C.

The overall reduction of the ore to metal occurs in a series of reactions, depending on the temperature range.

$$\sim 250°C \quad 3Fe_2O_3 + CO \rightarrow 2Fe_3O_4 + CO_2$$
$$\sim 500°C \quad Fe_3O_4 + CO \rightleftharpoons 3FeO + CO_2$$
$$\sim 700°C \quad FeO + CO \rightleftharpoons Fe + CO_2$$

In addition to these reactions the limestone is thermally decomposed and the carbon dioxide reacts with coke to form carbon monoxide

$$CaCO_3 \rightarrow CaO + CO_2$$
$$CO_2 + C \rightarrow 2CO$$

The former reaction produces lime which reacts with the silica impurities of the ore to form a molten slag, which floats at the surface of the molten iron and can be tapped off at the bottom of the furnace.

$$CaO + SiO_2 \rightarrow CaSiO_3$$

This slag may be used in the manufacture of cement (Chapter 7). The latter reaction helps to maintain the high concentration of carbon monoxide which is necessary for the reduction reactions in the upper zones of the furnace.

The metal obtained direct from the blast furnace is called *pig iron* and still contains a fair percentage of impurities, particularly carbon together with smaller amounts of manganese, silicon and phosphorus. This crude material is then further refined to form any one of a large number of possible products.

Cast Iron is produced by remelting and cooling pig iron, the nature of the product being determined mainly by the rate of cooling and by the chemical composition. The most important factor here is the way the carbon present is incorporated in the final material. The carbon may exist as flakes of graphite distributed throughout the metal, giving a relatively soft material called *grey cast iron*, or in the form of *cementite* (Fe_3C) producing a hard brittle metal described as *white cast iron*.

If molten pig iron is cooled quickly the cementite has no time to decompose and the resulting product is white cast iron. Slow cooling, on the other hand, assists the formation of graphite thus producing grey cast iron. Apart from carbon, cast iron contains relatively small amounts of silicon and sulphur, both of which affect the type of iron produced. The presence of silicon aids the formation of graphite so producing a grey iron, while sulphur has the effect of stabilizing the cementite thus producing a white iron.

Steel is manufactured by further refining the pig iron produced by the blast furnace. The aim is to remove the impurities either as gases or as slags and then to add carefully controlled amounts of, for example, carbon, chromium or manganese to give a steel with the required properties.

Cobalt and nickel tend to be used as alloys rather than in the pure state, although nickel itself has many properties similar to mild steel. Alloyed with aluminium and iron, cobalt and nickel form the material *alnico* which is widely used in the manufacture of permanent magnets. Nickel alloys can be made with the same coefficient of expansion as glass and can be used as 'seal-in' wires in, for example, electric light bulbs; while cobalt alloyed with iron is used in the manufacture of high-speed cutting tools and surgical instruments.

9.8. Group IB elements

The elements of group IB, copper, silver and gold are generally referred to as the *coinage metals*, and have long been used by man because of their pleasing appearance and resistance to corrosion. Table 9.7 shows that while we might expect some similarity to group IA metals in that they all possess a single outer valence electron, in fact the two groups are very dissimilar.

TABLE 9.7. THE COINAGE METALS

Element	Electronic configuration	Atomic radius (Å)	M.p. (°C)	B.p. (°C)
Cu	2.8.18.1	1·17	1083	2336
Ag	2.8.18.18.1	1·34	961	1950
Au	2.8.18.32.18.1	1·34	1063	2600

The group IA elements were light, low melting metals, and were extremely reactive. The coinage metals, on the other hand, are dense, high melting materials and chemically unreactive — so much so that they are found in the native state or as easily reducible ores.

The principal ores of copper are *chalcocite* (Cu_2S), *copper pyrites* ($CuFeS_2$) and *malachite* ($CuCO_3 \cdot Cu(OH)_2$), from which the metal is readily obtained by a simple smelting process, followed by electrolytic refining. Technically, copper probably comes second only to iron in importance. The pure metal is an excellent conductor of both heat and electricity and is consequently widely used in the electrical industry. It is also a very easily worked metal, with a very high resistance to corrosion. These properties make it a valuable constructional material, e.g. for chemical plant, boilers, roofing. Apart from the many uses for the pure metal, copper forms a large number of useful alloys, e.g. *brass* (Cu–Zn), *bronze* (Cu–Zn–Sn), *German silver* (Cu–Zn–Ni).

Because of its low electrode potential, copper does not dissolve in acids with the liberation of hydrogen gas, but will dissolve in oxidizing acids or in acids where it can form complex ions:

$$3Cu + 8HNO_3 \rightarrow 3Cu(NO_3)_2 + 4H_2O + 2NO$$

The most important salt is the sulphate which is used extensively as a fungicide and germicide, e.g. *Bordeaux mixture* is a mixture of copper sulphate and lime.

Silver occurs mainly as the native metal or as the ore *argentite* (Ag_2S). One method of extracting silver from its ores depends on the formation of soluble silver complexes with cyanide. Thus either crude silver or the sulphide ore can be made to dissolve in a sodium cyanide solution through which air is blown:

$$4Ag + 8NaCN + O_2 + 2H_2O \rightarrow 4Na[Ag(CN)_2] + 4NaOH$$
$$2Ag_2S + 8NaCN + O_2 + 2H_2O \rightarrow$$
$$4Na[Ag(CN)_2] + 4NaOH + 2S$$

The solution containing the complex silver ions may then be reduced with zinc or aluminium to yield metallic silver:

$$Zn + 2Na[Ag(CN)_2] \rightarrow Zn(CN)_2 + 2NaCN + 2Ag$$

The metal has a very high lustre, is easily worked, and is the best known conductor of heat and electricity. Although the latter properties are not generally utilized because the metal is expensive, silver is widely used as a decorative metal because of its attractive appearance. It does not corrode readily except in the presence of sulphur-containing compounds where it tarnishes fairly readily due to the formation of black silver sulphide:

$$4Ag + 2H_2S + O_2 \rightarrow 2Ag_2S + 2H_2O$$

Silver is a very soft metal and normally is used as an alloy (e.g. with copper) or as a thin coating plated on a base metal. This can readily be achieved by electroplating, whereby the article to be coated is made the cathode of an electrolytic cell, with pure silver as the anode, and a solution of the silver cyanide complex ($NaAg(CN)_2$) as electrolyte.

One interesting application of silver salts is in photography, where use is made of the light sensitivity of certain silver salts, and of their ability to form complexes. The photographic film is a plastic or glass plate coated with an emulsion of silver bromide in gelatin. When the film is exposed the silver bromide granules are affected by the light and a small amount of reduction takes place, with the formation of metallic silver. Development of the film is simply a process of reduction, where a chemical reducing agent (e.g. pyrogallol) is used to accentuate the reduction process which occurred on exposure. The areas of the plate subjected to intense light are rapidly reduced, whereas areas not subjected to the same extent reduce much more slowly. The fixing process is one which removes the silver bromide remaining unchanged on the plate. This is dissolved off by treatment with sodium thiosulphate ('hypo') giving a plate which may then be safely exposed to light.

$$3\,AgBr + 3\,Na_2S_2O_3 \rightarrow 2\,NaBr + Na_4[Ag_2(S_2O_3)_3]$$

9.9.   Group IIB elements

Although the metals of this group, zinc, cadmium and mercury, each possess two outer valence electrons, they resemble the transition metals rather more closely than they do the group IIA metals, which have similar electronic configurations.

TABLE 9.8. THE GROUP IIB METALS

Element	Electronic configuration	Atomic radius (Å)	M.p. (°C)	B.p. (°C)
Zn	2.8.18.2	1·25	419	907
Cd	2.8.18.18.2	1·41	321	767
Hg	2.8.18.32.18.2	1·44	−38·9	357

The principal zinc ore is *zinc blende* (ZnS) from which the zinc may readily be obtained. After a roasting process to convert the sulphide to oxide, the material is reduced with coke to form the metal which is then simply distilled out of the furnace.

Although zinc is chemically a fairly reactive metal it is widely used as a protective coating for iron because of its resistance to atmospheric corrosion due to protective film formation, and because it is anodic to iron. Zinc is also widely used in the formation of alloys, e.g. with copper to give *brass*, or with small amounts of aluminium to give useful die-casting alloys. Both zinc oxide and sulphide are used as white pigments, while the sulphide, which acts as a phosphor, is used for making fluorescent screens.

Cadmium is chemically very similar to zinc and the two elements are usually found together. Like zinc, cadmium may be used as a protective coating and is used in the formation of bearing alloys with a low coefficient of friction.

Mercury, the only metal which is a liquid at room temperature, is occasionally found in the native state but more generally as the ore *cinnabar* (HgS). The metal may be obtained simply by roasting, since the oxide initially formed is thermally unstable and decomposes:

$$HgS + O_2 \rightarrow Hg + SO_2$$

The metal is chemically inert, has a high density, but is for a metal a poor conductor of electricity. It is nevertheless used in the manufacture of scientific equipment, e.g. thermometers, barometers, vacuum pumps, and in general for making electrical contacts where its mobility outweighs its poor conductivity. A very large number of metals dissolve in mercury to form *amalgams*, which may be either liquid or solid.

Mercury compounds, particularly organometallics, are widely used as fungicides and herbicides, e.g. the oxide is used as an antifouling paint. Mercury and its soluble compounds are all highly toxic and great care must be taken in their use.

CHAPTER 10

Solutions

A SOLUTION is a homogeneous dispersion of two or more substances. The substance dispersed is called the *solute* whilst the
dispersion medium is the *solvent*. In true solutions the solute is
dispersed as individual molecules or ions. Common examples
of solute-solvent pairs are:

gas–liquid	e.g. air in water
liquid–liquid	e.g. ethanol in water
solid–liquid	e.g. salt in water.

Of the above three types, the most important are liquid–liquid
and solid–liquid solutions and in this chapter some of the more
important properties of such solutions will be discussed.

10.1. Solid–liquid solutions – vapour pressure

When a non-volatile solute is dissolved in a suitable solvent
the vapour pressure of the solvent is lowered. This is the result
we would expect for in a solution the solvent molecules experience not only a mutual attraction for each other but also an
attraction for the solute molecules. (If no such attraction existed
then the solute would not have dissolved in the solvent in the
first place.) As a result, the 'escaping tendency' of the surface
solvent molecules is lowered. On this basis, we could reasonably predict that the magnitude of the effect would depend on
the solute concentration. This dependence is described by
Raoult's law which states that *the relative lowering of the
vapour pressure is equal to the mole fraction of solute in the*

160

solution. Thus, if $P°$ and P are the vapour pressures of pure solvent and solution (both solvent and solution being at the same temperature) then

$$\frac{P° - P}{P°} = \frac{\Delta P}{P°} = x_2$$

where x_2 = mole fraction of solute,
          ΔP = actual lowering of vapour pressure,
          $\Delta P/P°$ = relative lowering of vapour pressure.

A solution which accurately obeys Raoult's law at all concentrations is said to be an ideal solution. Such ideal behaviour will manifest itself provided the attractive interactions between the two different molecular species are of the same magnitude as those between the similar species. In reality, deviations from ideal behaviour can arise, but if the solution is dilute, such deviations are usually (but not always) sufficiently small to be ignored.

For a particular solvent at constant temperature, $P°$ is a constant. Hence the magnitude of the vapour pressure lowering, ΔP, will depend exclusively on the concentration of the solute (i.e. the number of molecules or ions in a given amount of solvent) but not at all on the nature of the solute.

Now, although the value of ΔP is not influenced by the nature of the solute, its value is dependent on the nature of the solvent. Thus, for a given solute concentration the value of ΔP will depend on the vapour pressure of the pure solvent, i.e. $\Delta P \propto P°$ when temperature and solute concentration x_2 are constant.

Properties of solutions which exhibit the above features are termed *colligative properties*.

10.2. Solid–liquid solutions – boiling point

A direct consequence of Raoult's law is that the boiling point of a pure liquid is elevated by the addition to it of a non-volatile solute. In Fig. 10.1 the vapour pressure curve of solvent is compared with the corresponding curve for the solution. At

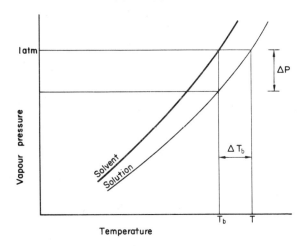

Fig. 10.1. Variation of vapour pressure with temperature for both
solvent and solution.

temperature T_b, the vapour pressure of the pure solvent is 1
atm. Consequently, if the liquid is heated under a constant
pressure of 1 atm, it will boil at this temperature. However, the
solution at temperature T_b exerts a vapour pressure less than
1 atm and if it is to be brought to the boil its temperature must
be raised to a value at which there is once more equality between vapour pressure and external pressure. This condition is
satisfied at temperature T.

Experimentally it is found that the elevation of the boiling
point, ΔT_b, is directly proportional to the lowering of the vapour
pressure, ΔP, which has already been shown to be proportional
to the solute concentration expressed by x_2.

Thus $\Delta T_b \propto x_2 = kx_2$

It is customary to relate the boiling-point elevation to the
solute concentration in terms of *molality* (m) rather than mole
fraction, where the molality is defined as the number of moles
of solute dissolved in 1000 grams of solvent. This substitution
of molality for mole fraction is only justifiable in the case of a

dilute solution. To illustrate this, let us consider w_2g of solute dissolved in w_1g of solvent. It follows, by simple proportion, that the weight of solute contained in 1000 g of solvent is $1000w_2/w_1$.

$\therefore$ Moles of solute in 1000 g of solvent $= \dfrac{1000w_2}{w_1M_2}$

where M_2 is the molecular weight of solute.

This quantity is, by definition, the molality (m) of the solution, i.e.

$$m = \frac{1000w_2}{w_1M_2}$$

Multiplication of both sides of the above equation by M_1, the molecular weight of the solvent, yields

$$mM_1 = \frac{1000w_2M_1}{w_1M_2} = \frac{1000n_2}{n_1}$$

where n_1 and n_2 are the moles of solute and solvent respectively.

If the solution is dilute then

$$\frac{n_2}{n_1} \approx x_2, \quad \text{where } x_2 = \frac{n_2}{n_1 + n_2} \quad \text{and } n_1 \gg n_2$$

$\therefore$ For a *dilute* solution $x_2 = \dfrac{mM_1}{1000}$

Substitution of this value in the equation

$$\Delta T_b = kx_2$$

yields

$$\Delta T_b = \frac{kM_1 \cdot m}{1000}$$

As k and M_1 are constants characteristic of the solvent, they can be combined to give

$$\Delta T_b = K_b \cdot m \quad \text{where } K_b = \frac{kM_1}{1000}$$

This new constant K_b, referred to as the *molal elevation constant*, is equivalent to the boiling-point elevation of a one molal solution.

EXAMPLE 10.1. *Benzene has a normal boiling point of* 80·2°C. *When* 3·0 g *of a non-volatile substance of molecular weight* 82 *is dissolved in* 100 g *of benzene the solution is allowed to boil until a temperature of* 81·7°C *is reached. What weight of benzene is removed as vapour?* K_b *for benzene* = 2·53.

When the above solution boils, only benzene vapour is lost; therefore, as boiling continues the solution becomes progressively more concentrated and boils at a progressively increasing temperature, so that when $\Delta T_b = 1.5$,

$$m = \frac{\Delta T_b}{K_b} = \frac{1·5}{2·53} = 0·593$$

∴ Solute concentration at 81·7°C = 0·593 mole per 1000 g of benzene
$\equiv$ 0·593 × 82 g per 1000 g of benzene
= 3 g per 61·7 g of benzene
wt. of solvent lost = 100 − 61·7 = 38·3 g

10.3. Solid–liquid solutions — freezing point

Reference to Fig. 10.2 shows that the freezing point of a solution is depressed below that of the pure solvent. Such is always the case provided only the solvent crystallizes out from solution.

The depression ΔT_f is, like ΔT_b, found to be proportional to the lowering of vapour pressure.

Thus $\Delta T_f \propto \Delta P \propto x_2$ and $\Delta T_f = k' x_2$

where k' has, as before, a value specific to the solvent.

$$\Delta T_f = \frac{k' M_1}{1000} m = K_f m \quad \text{where} \quad K_f = \frac{k' M_1}{1000}$$

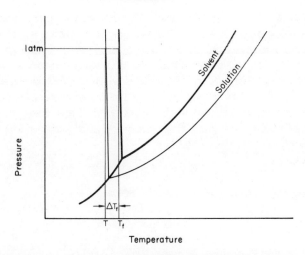

Fig. 10.2. Influence of pressure on the freezing point of both solvent and solution.

The constant K_f is called the *molal depression constant*, and is equivalent to the freezing point depression for a one molal solution.

Both boiling-point elevation and freezing-point depression are, like vapour pressure lowering, colligative properties, in that the magnitude of the property is determined not by the nature of the solute but by its concentration, the latter being expressed in terms of the molality of the solution. Also, for a given solute concentration both ΔT_b and ΔT_f will depend on the value of K_b and K_f respectively, i.e. on the nature of the solvent.

EXAMPLE 10.2. *0.2 g of calcium chloride is dissolved in* 100 *g of water. At what temperature will the resulting solution freeze?* K_f *for water* $= 1.86$.

Molality of calcium chloride $= \dfrac{\text{wt.}}{\text{mol. wt.}} \times \dfrac{1000}{100} = \dfrac{0.2}{111} \times \dfrac{1000}{100}$
$$= 0.018$$

At first sight we would expect the depression in the freezing point to be $1·86 \times 0·018 = 0·03348°$. In fact, such a solution as the one above would give a depression substantially greater than $0·03348°$. To understand why this should be so we must bear in mind that calcium chloride is an ionic substance and the process of solution is one in which the water molecules separate ions already present in the solid state. Thus, for every $1·0$ mole of $CaCl_2$ dissolved, there are present in solution $1·0$ g ion of Ca^{++} and $2·0$ g ions of Cl^-.

$$CaCl_2 \rightarrow Ca^{++} + 2Cl^-$$

Hence the *actual* molality of the solution

$$= \text{molality of } Ca^{++} + \text{molality of } Cl^-$$
$$= 0·018 + 0·036$$
$$= 0·054$$

$\therefore \Delta T_f = 1·86 \times 0·054 = 0·1004° (= 3 \times 0·03348)$

$\therefore$ The solution will freeze at $-0·1004°C$.

Even this answer is only approximate; in reality the observed depression would be slightly less than $0·1004°$ on account of interionic attraction between the positive and negative ions. Such attraction prevents the ions from acting completely independently of each other. However, this effect is influenced by the distance separating the ions. Thus if a solution is of high concentration the ions will, on average, be near each other and interionic attraction will be strong. In a dilute solution, however, the ions will be separated by much larger distances; interionic effects will then be weak, and as a result, the calculated depression will be very nearly equal to the observed value.

10.4.   Liquid–liquid solutions — vapour pressure

Up till now, the solute has been considered to be non-volatile, and our only concern has been the influence of solute on the vapour pressure of solvent; this was expressed by the equation

$$\frac{\Delta P}{P^{\circ}} = x_2$$

For our present purpose, it is convenient to alter the form of this relationship. Thus

$$\frac{\Delta P}{P^\circ} = \frac{P^\circ - P}{P^\circ} = 1 - \frac{P}{P^\circ} = x_2$$

But $x_1 + x_2 = 1$, where x_1 = mole fraction of solvent

∴ $$P = x_1 P^\circ$$

At constant temperature P° is a constant

∴ $$P \propto x_1$$

i.e. *the vapour pressure of solvent in solution is directly proportional to the mole fraction of solvent in the solution.* This statement is often quoted as an alternative form of Raoult's law.

If a solution is made up of two volatile components,† e.g. two liquids A and B, a complication is introduced, for we must now consider not only the influence that B has on the vapour pressure of A but also the effect that A has on the vapour pressure of B. This added complication is most easily dealt with if we imagine for the moment that liquid B is non-volatile. In this event the vapour pressure of liquid A in solution is related in the ideal case to the mole fraction of A by the equation:

$$P_A = x_A \cdot P_A^\circ$$

where P_A° = vapour pressure of pure A,

P_A = ~~partial~~ vapour pressure of A in solution,

x_A = mole fraction of A in solution.

We must now establish the contribution made by liquid B to the total pressure. This is done by imagining liquid A to be the non-volatile component. Thus if

P_B° = vapour pressure of B,

P_B = ~~partial~~ vapour pressure of B in solution,

x_B = mole fraction of B in solution,

then $P_B = x_B \cdot P_B^\circ$ in the ideal case

†For such a solution, the terms solute and solvent cease to have any very obvious meaning. In this event, the material present in excess is normally called the solvent and the other the solute.

Applying Dalton's law of partial pressures, it follows that

$$P_{\text{total}} = P_A + P_B = x_A . P_A^{\,\circ} + x_B . P_B^{\,\circ}$$

This equation reveals the way in which the total pressure varies with the composition of the solution. The relationship is made clearer when $1 - x_A$ is substituted for x_B.

Thus
$$P_{\text{total}} = x_A . P_A^{\,\circ} + (1 - x_A)P_B^{\,\circ}$$
$$- (P_A^{\,\circ} - P_B^{\,\circ})x_A + P_B^{\,\circ}$$

At constant temperature, $P_A^{\,\circ}$ and $P_B^{\,\circ}$ are themselves constant and so the equation in this form is the equation of a straight line when P_{total} is plotted against x_A (Fig. 10.3) with the gradient $= P_A^{\,\circ} - P_B^{\,\circ}$ and the intercept $= P_B^{\,\circ}$. This line is called the *liquid composition line (L)*. Also shown in Fig. 10.3 are the individual contributions made to the total pressure by A and B separately. Thus, *for ideal liquid–liquid solutions, the vapour pressure is always intermediate between the vapour pressures of the pure components.*

If we continue our analysis of this type of system, it is possible to show that the vapour produced from and in equilibrium

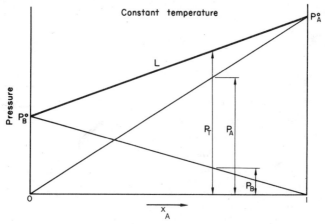

Fig. 10.3. Variation of vapour pressure with liquid composition.

with an ideal liquid mixture is not of the same composition as the original liquid mixture but in fact is always richer in the more volatile component. This fact can be demonstrated in the following manner:

$$P_{\text{total}} = x_A P_A{}^\circ + x_B P_B{}^\circ$$

From Dalton's law $P_A = x_A{}' \cdot P_{\text{total}}$ where $x_A{}'$ is the mole fraction of A in the *vapour*,

i.e. $$P_{\text{total}} = \frac{P_A}{x_A{}'}$$

$\therefore \qquad \dfrac{P_A}{x_A{}'} = x_A \cdot P_A{}^\circ + x_B \cdot P_B{}^\circ$

$\therefore \qquad \dfrac{x_A P_A{}^\circ}{x_A{}'} = x_A \cdot P_A{}^\circ + x_B \cdot P_B{}^\circ \qquad P_A = x_A \cdot P_A{}^\circ$

$\therefore \qquad \dfrac{x_A(\text{mole fraction of A in the liquid})}{x_A{}'(\text{mole fraction of A in the vapour})} = x_A + x_B \cdot \dfrac{P_B{}^\circ}{P_A{}^\circ}$

Now $x_A + x_B = 1$. In the event of $P_B{}^\circ$ being equal to $P_A{}^\circ$ (at one particular temperature), x_A will equal $x_A{}'$. Such a condition is not impossible but it is distinctly improbable. If, on the other hand, $P_A{}^\circ$ is greater than $P_B{}^\circ$ then x_A' must be less than x_A, i.e. *the vapour contains a higher proportion of A, the more volatile component, than does the original liquid mixture.*

EXAMPLE 10.3. *A solution of 0·6 mole fraction heptane and 0·4 mole fraction octane boils at 40°C at a pressure of 67·3 mm Hg. At this temperature, the vapour pressure of pure heptane is 91·5 mm. What is the composition of the vapour initially produced from the boiling liquid?*

The liquid mixture boils at 40°C under a pressure of 67·3 mm Hg. It therefore follows that the total pressure exerted by the solution at 40°C is also equal to 67·3 mm Hg.

Ideally $$P_{\text{total}} = x_H P_H{}^\circ + x_0 P_0{}^\circ$$

where x_H and x_0 are the mole fractions of heptane and octane in the solution, and P_H° and P_0° are the vapour pressures of pure heptane and octane respectively.

$$67\cdot3 = 0\cdot6 \times 91\cdot5 + 0\cdot4 P_0^\circ$$
$$P_0^\circ = 31 \text{ mm}$$

we now have all the information required to compute the initial vapour composition using the equation

$$\frac{x_H \text{ (mole fraction of heptane in the liquid)}}{x_H' \text{ (mole fraction of heptane in the vapour)}} = x_H + x_0 \cdot \frac{P_0^\circ}{P_H^\circ}$$

$$\therefore \qquad\qquad x_H' = 0\cdot8158$$

If values for P_A° and P_B°, x_A and x_B are known and substituted in the above equation, corresponding x_A' values can be obtained. These can then be plotted against the total pressure (Fig. 10.4) to yield a curve called the *vapour composition curve (V)*.

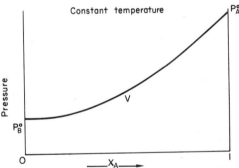

FIG. 10.4. Variation of vapour pressure with vapour composition.

The results expressed graphically in Figs. 10.3 and 10.4 are combined in Fig. 10.5.

10.5.   Liquid–liquid solutions — boiling point

Fig. 10.5 represents the liquid–vapour equilibrium by means of a pressure-composition diagram. An alternative description can be used in which the boiling point is plotted against the

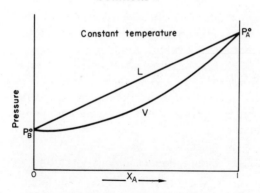

FIG. 10.5. Liquid and vapour composition curves.

composition (Fig. 10.6). Thus if A is the more volatile component, $P_A° > P_B°$ and $T_A < T_B$, where T_A and T_B are the boiling points of pure A and B. The lower curve (L) describes the liquid composition, while the upper curve (V) describes the vapour composition. Thus, a liquid mixture of composition l will boil at temperature T to produce an initial vapour of composition v. This vapour is seen to be richer in component A, the more volatile component.

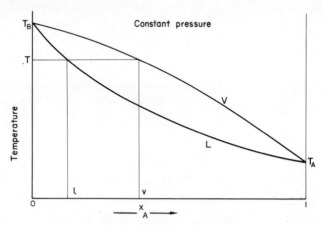

FIG. 10.6. Boiling point composition diagram.

The type of solution so far discussed is an ideal one, i.e. one in which each component is considered to obey Raoult's law at all concentrations. As with solid–liquid solutions, few liquid-solutions behave completely ideally and very marked deviations from ideality are not at all uncommon. A lengthy discussion of such systems is beyond the scope of this book, although reference is made to non-ideal liquid–liquid systems in Chapter 16.

10.6. Liquid–liquid solutions — freezing point

In Section 10.3 we saw that the freezing point depression could be thought of as being a consequence of vapour pressure lowering. While such a treatment is perfectly valid for solid–liquid solutions, it is inapplicable to liquid–liquid systems, which accordingly must be treated differently. However, it is found that, provided only one component crystallizes out from solution, the same relationship holds good, viz.

$$\Delta T_f = K_f \cdot m$$

10.7. Antifreezes

In the internal combustion engine, only a portion of the energy released in the combustion of gasoline is converted into power; the remainder is dissipated as heat. Part of this heat is removed in the exhaust gases whilst the remainder is transferred to the cooling system, in which (in most cars) water is caused to circulate through the engine block. Water, however, freezing at 0°C, can easily 'ice-up' during the winter months and cause the block to crack. To overcome this problem a variety of substances called *antifreezes* can be added to the water to reduce its freezing point to a level below that of temperatures likely to be encountered during the winter months.

The two most popular antifreeze compounds in use at present are methyl alcohol and ethylene glycol, although ethyl alcohol, isopropyl alcohol, glycerol and propylene glycol have some properties which make them moderately suitable as antifreeze agents. On a weight to weight basis the efficiency of these materials, in their ability to lower the freezing point, is related to the molecular weight of the antifreeze.

Thus if

$$\Delta T_f = K_f . m$$

Then

$$\Delta T_f = \frac{1000K_f . w_2}{M_2 w_1}$$

Now if the ratio w_2/w_1 is made the same for each antifreeze in turn, then inspection of the above equation shows that the depression is inversely proportional to the molecular weight of antifreeze,

i.e.

$$\Delta T_f \propto 1/M_2$$

As the relationship has been developed from equations which strictly apply only to ideal, dilute solutions, it should be applied with care. However, if two materials have markedly different molecular weights, it is possible to assess their relative efficiencies in functioning as antifreezes. Thus, on this basis alone, methyl alcohol (mol. wt. = 32) is clearly a more efficient antifreeze than ethylene glycol (mol. wt. = 62) which suffers from the further disadvantage of being dearer than the very cheaply available methyl alcohol. However, it too has its disadvantages, viz. its greater tendency to give corrosive products, the toxicity of its fumes and its high volatility. Ethylene glycol in these respects is the more suitable antifreeze and overall is favoured more than methyl alcohol. Its low volatility is a particularly strong argument in favour of its use, for when a solution of ethylene glycol (b.p. 197°C) in water boils, it does so at a temperature in excess of 100°C, whereas a solution of methyl alcohol (b.p. 64°C) in water boils at a temperature intermediate between 64°C and 100°C. Now when a car engine is turned off, the engine transfers some of its heat to the antifreeze in the cooling system thereby raising its temperature. If methyl alcohol is used, this increase in temperature may cause some of the antifreeze to 'boil off'. Loss of radiator fluid does not occur if a relatively non-volatile material such as ethylene glycol is used.

Irrespective of what antifreeze is used, extensive corrosion will occur in the cooling system. To counteract this, *inhibitors* (Section 8.11) are added to reduce the extent of corrosion to a tolerable level.

EXAMPLE 10.4. *Calculate the freezing point of a methyl alcohol–water solution containing 15 per cent by weight of methyl alcohol. What weight per cent of ethylene glycol would be required to give the same depression in the freezing point? K_f for water = 1·86.*

Every 100 g of this solution contains 15 g of methyl alcohol and 85 g of water.

$$\therefore \qquad \text{Molality of solution} = \frac{15 \times 1000}{32 \times 85} = 5\cdot516$$

$$\therefore \qquad \Delta T_f = 1\cdot86 \times 5\cdot516 = 10\cdot26°$$

$$\therefore \qquad \text{Freezing point of solution} = -10\cdot26°C$$

If ethylene glycol is to afford the same protection, the molality of the ethylene glycol–water solution must be as before, viz. 5·516,

i.e.   Ethylene glycol concentration

$$= 5\cdot516 \text{ mole per 1000 g of water}$$
$$\equiv 5\cdot516 \times 62 \text{ g per 1000 g of water}$$
$$= 342\cdot1 \text{ g per 1000 g of water}$$

$$\text{Per cent by weight of ethylene glycol} = \frac{342\cdot1}{1342\cdot1} \times 100$$

$$= 25\cdot49 \text{ per cent.}$$

EXAMPLE 10.5. *What weight of ice would crystallize out if 100 g of the ethylene glycol solution of Example 4 was cooled to –15°C?*

When the above solution is cooled it begins to freeze at –10·26°C; i.e. ice crystals appear for the first time at this temperature. As the temperature is further reduced more ice separates out, the solution becomes more concentrated and the freezing point is progressively depressed. It is thus seen that an aqueous solution, unlike pure water, does not freeze solid when its temperature is reduced below its freezing point; instead it becomes progressively more concentrated as more and more ice separates out. Now as the solid phase separating out is pure water it follows that, as the temperature drops from $-10\cdot26°C$ to $-15°C$, the solution must throughout contain the initial amount of ethylene glycol. By comparing the initial glycol concentration with the glycol concentration of the solution at $-15°C$ we can deduce the weight of ice separating out. Thus:

Initial glycol concentration $= 25\cdot49$ g per $74\cdot51$ g of water.

When $\Delta T = 15$, $m = \dfrac{15}{1\cdot86} = 8\cdot065$

$\therefore$ Glycol concentration at $-15°C$

$= 8\cdot065$ mole per 1000 g of water
$\equiv 500$ g per 1000 g of water
$= 25\cdot49$ g per $51\cdot98$ g of water

$\therefore$ Wt. of ice separating out $= 74\cdot51 - 51\cdot98$
$= 22\cdot53$ g

Ultimately a stage is reached when the system is entirely solid but this occurs at a temperature well below that at which ice crystals first appear. It is thus seen that the addition of an antifreeze to the water of the cooling system provides a high degree of protection against "icing-up".

CHAPTER 11

Thermochemistry

11.1. Energy of a chemical system

The energy content of any system is partly kinetic, partly potential. A gas, for example, will possess kinetic energy by virtue of the chaotic motion of the molecules (the amount of kinetic energy being dependent on the temperature); and it will possess potential energy as a result of a particular arrangement of protons, neutrons and electrons within the molecules (this arrangement being specific to a particular substance). In short, its energy content will depend on the state of the system, i.e. on the temperature, pressure, volume, amount and nature of the gas in question. The same is true for liquids and solids.

A measure of this energy content is a quantity referred to as the *enthalpy (H)* of the system.† This quantity, whose value is a function of the state of the system, is sensibly called a *state function*.

There are two important consequences of the fact that enthalpy is a state function:

(i) its value will not depend on the way in which the state has been reached;

(ii) when the system experiences some change in its state (e.g. as a result of reaction occurring when reactant material is replaced by product material), the change in enthalpy (ΔH) will depend only on the initial and final states of the system,

†An absolute value for H cannot be obtained. Fortunately, however, this is no cause for concern, for the chemist is interested primarily in a change in energy content, i.e. a change in enthalpy.

i.e. $$\Delta H = H_{\text{final}} - H_{\text{initial}}$$

A system which provides a useful analogy to explain the nature of a state function is illustrated in Fig. 11.1, where a ball is pictured as lowering its position by x cm, in the first case by rolling down an incline, in the second case by dropping vertically. In each case, the change in potential energy is the same, i.e. the potential energy of the ball depends only on its position, or *state*, and not on the route by which it has reached that position. The potential energy of the ball is a state function, and any change in this quantity will be given simply by the difference between the initial and final values of the function.

i.e. $$\Delta\text{P.E.} = \text{P.E.}_{\text{final}} - \text{P.E.}_{\text{initial}}$$

Both the distance travelled and the time taken for the ball to change its position are route dependent – these are not state functions.

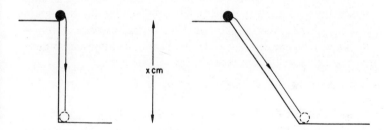

Fig. 11.1. The energy of a ball as a function of its position.

11.2. Exothermic and endothermic reactions

Consider the general reaction

$$a\text{A} + b\text{B} \rightarrow c\text{C} + d\text{D}$$

to take place under conditions of constant pressure and temperature. Then we can say that

$$\text{enthalpy of reactants} = aH_A + bH_B = H_{\text{reactants}} = H_R$$

and
$$\text{enthalpy of products} = cH_C + aH_D = H_{\text{products}} = H_P$$

where H_A, H_B, H_C, H_D are the enthalpies per mole of materials A, B, C and D.

$$\therefore \quad \text{Change in enthalpy} = \Delta H = cH_C + dH_D - (aH_A + bH_B)$$
$$= H_P - H_R$$

Suppose now that H_R is greater than H_P by an amount x cal. The principle of energy conservation demands that the *total* energy of the system remain constant. An energy balance would therefore require that

$$H_R = H_P + x$$

The quantity x cal is energy liberated as heat.

Now $\qquad\qquad H_P - H_R = \Delta H = -x$ cal.

Hence a change in enthalpy is seen to be identical in magnitude to the quantity of heat liberated under the conditions of constant pressure and temperature. It is, however, of opposite algebraic sign. For this reason, the heat change accompanying reaction can be expressed in one of two ways:

(i) $aA + bB \rightarrow cC + dD + x$ cal

or

(ii) $aA + bB \rightarrow cC + dD \qquad \Delta H = -x$ cal

The convention in chemistry (which will be used in this text) is to view heat changes from the point of view of the reacting system. A system which loses heat is therefore given a negative value of ΔH, and vice versa. Reactions characterized by a negative value of ΔH are described as being *exothermic*. Conversely, reactions for which ΔH is positive, $(H_P > H_R)$ are accompanied by an absorption of heat and are referred to as being *endothermic*. For example, if the above reaction could be reversed, x cal would be absorbed and ΔH would equal $+x$ cal

$$cC + dD \rightarrow aA + bB - x \text{ cal} \qquad \Delta H = x \text{ cal}$$

We thus see that any chemical reaction which is exothermic in a forward direction will be endothermic to precisely the same extent if reversed.

11.3. Heat of formation

It is of considerable interest to the chemist to know the degree to which a reaction is exothermic or endothermic. This requires a knowledge of ΔH. For a number of reasons it is not always possible to obtain experimentally a value of ΔH; indeed it would be extremely laborious to perform an experiment each time a ΔH value was required. One method of calculating ΔH which might occur to the reader would be to use the equation

$$\Delta H = H_P - H_R$$

If the enthalpies of all the materials involved in the reaction were in fact known, it would be no more than a simple arithmetical exercise to calculate ΔH. However, as absolute values of enthalpies are not known, an alternative approach which makes use of *heats of formation* is used.

The heat of formation of a compound is the change in enthalpy accompanying the formation of one mole of the compound from its constituent elements in their standard states in which they usually occur under normal conditions of pressure and temperature. It is given the symbol ΔH_f, and is measured at 25°C (298°K) and 1 atm.

To avoid ambiguity it is wise in thermochemical equations to use the letters s, l, or g to denote the physical state of each substance. Thus the heat of formation of liquid water is indicated by

$$H_2(g) + \tfrac{1}{2}O_2(g) \rightarrow H_2O(l) \qquad \Delta H_f = -68 \cdot 32 \text{ kcal}$$

while the heat of formation of water vapour is indicated by

$$H_2(g) + \tfrac{1}{2}O_2(g) \rightarrow H_2O(g) \qquad \Delta H_f = -57 \cdot 8 \text{ kcal}$$

As a consequence of the way in which heat of formation is defined, it follows that the heat of formation of elements in their standard states is zero. For example, the heat of formation of gaseous oxygen is, by definition, the enthalpy change accompanying the formation of one mole of gaseous oxygen from the state in which it normally occurs, viz. the gaseous state,

i.e. $$\Delta H_{f(O_2, g)} = 0$$

Generalizing, $\Delta H_{f(\text{elements in their standard states})} = 0$

Table 11.1 lists the heats of formation of a variety of compounds at 298°K.

TABLE 11.1. HEATS OF FORMATION, ΔH_f, kcal mole^{-1} AT 298°K

$H_2O(l)$	-68.32	$C_2H_4(g)$	12.5
$H_2O(g)$	-57.8	$CH_3OH(l)$	-57.02
$CO(g)$	-26.42	$C_2H_5OH(l)$	-66.36
$CO_2(g)$	-94.1	$SO_2(g)$	-70.96
C(graphite)	0	$SO_3(g)$	-94.45
C(diamond)	-0.453	$HCl(g)$	-22.06
$CH_4(g)$	-17.89	$HBr(g)$	-8.66
$C_2H_6(g)$	-20.24	$HI(g)$	6.2

11.4. Heat of reaction

In what way are the heats of formation of compounds involved in a chemical reaction related to the overall heat of reaction?

Consider the following reaction in which carbon disulphide is burned in oxygen to yield carbon dioxide and sulphur dioxide. Let the enthalpy change be ΔH_1

$$CS_2(l) + 3O_2(g) \rightarrow CO_2(g) + 2SO_2(g) \qquad \Delta H_1 = -264.66 \text{ kcal}$$

Imagine now the reaction to proceed by an alternative route. Let the first step be the conversion of reactants to their elements in their standard states.

$$CS_2(l) \rightarrow C(s) + 2S(s) \qquad \Delta H_2 = -28.64 \text{ kcal}$$

where

$$\Delta H_2 = H_{(C)} + 2H_{(S)} - H_{(CS_2)} = -\Delta H_{f(CS_2)}$$

Let the second step be recombination of the elements to form products:

$$C(s) + O_2(g) \rightarrow CO_2(g) \qquad \Delta H_3 = -94.1 \text{ kcal}$$

where $$\Delta H_3 = \Delta H_{f(CO_2)}$$

$$2S(s) + 2O_2(g) \rightarrow 2SO_2(g) \qquad \Delta H_4 = -141.92 \text{ kcal}$$

where

$$\Delta H_4 = 2\Delta H_{f(SO_2)}$$

The overall enthalpy change $= \Delta H_2 + \Delta H_3 + \Delta H_4 = -264.66$ kcal

Thus the overall enthalpy change for the above sequence of reactions is seen to be equivalent to the enthalpy change for the direct reaction between carbon disulphide and oxygen. Such an equality follows as a consequence of enthalpy being a state function.

Thus

$$
\begin{aligned}
\Delta H_1 &= \Delta H_2 + \Delta H_3 + \Delta H_4 \\
&= -\Delta H_{f(CS_2)} + \Delta H_{f(CO_2)} + 2\Delta H_{f(SO_2)} \\
&= \Delta H_{f(CO_2)} + 2\Delta H_{f(SO_2)} - \Delta H_{f(CS_2)} \\
&= \Delta H_{f(\text{products})} - \Delta H_{f(\text{reactants})} \qquad (\Delta H_{f(O_2,g)} = 0)
\end{aligned}
$$

The reaction considered is one of combustion; the enthalpy change in a combustion reaction is referred to as the *heat of combustion*, and is defined as *the enthalpy change accompanying the complete oxidation of one mole of a compound*.

For *any* reaction the overall enthalpy change ΔH is related to the heats of formation of the various compounds involved by the simple equation

$$\text{heat of reaction, } \Delta H = \Delta H_{f(\text{products})} - \Delta H_{f(\text{reactants})}$$

This important relationship may be used in one of two ways.

(i) Knowing the heats of formation, the heat of reaction can be rapidly calculated, thus avoiding needless experimentation.

EXAMPLE 11.1. *The hydrocarbon 1,3-butadiene, used in the manufacture of synthetic rubber (Section 28.9), is made by the dehydrogenation of 1-butene*:

$$CH_2 = CH \cdot CH_2 \cdot CH_3(g) \rightarrow CH_2 = CH \cdot CH = CH_2(g) + H_2(g)$$

Determine the enthalpy change (at 25°C) in this reaction. The heats of combustion of 1-butene and 1,3-butadiene are, respectively, -649.8 and -607.9 kcal mole^{-1} at 25°C. The heat of formation of $H_2O(l)$ is -68.3 kcal mole^{-1} at 25°C.

For the dehydrogenation

$$\Delta H = \Delta H_{f(\text{products})} - \Delta H_{f(\text{reactants})}$$
$$= \Delta H_{f(H_2)} + \Delta H_{f(1,3\text{-butadiene})} - \Delta H_{f(1\text{-butene})} \quad (1)$$

Values for the respective heats of formation have not been given — this is not uncommon. They can, however, be deduced, in this case from information regarding heats of combustion as follows.

The combustion of 1-butene (a) and 1,3-butadiene (b) yields as products carbon dioxide and water.

(a) $CH_2 = CH \cdot CH_2 \cdot CH_3(g) + 6O_2(g) \rightarrow 4CO_2(g) + 4H_2O(l)$
$$\Delta H_1 = -649.8 \text{ kcal mole}^{-1}$$

Now $\Delta H_1 = 4\Delta H_{f(CO_2)} + 4\Delta H_{f(H_2O, l)} - \Delta H_{f(1\text{-butene})} - 6\Delta H_{f(O_2, g)}$

$\therefore \Delta H_{f(1\text{-butene})} = 4\Delta H_{f(CO_2)} + 4\Delta H_{f(H_2O, l)} - \Delta H_1 \quad \Delta H_{f(O_2, g)} = 0$

(b) $CH_2 = CH \cdot CH$
$$= CH_2(g) + 11/2 O_2(g) \rightarrow 4CO_2(g) + 3H_2O(l)$$
$$\Delta H_2 = -607.9 \text{ kcal mole}^{-1}$$

Now $\Delta H_2 = 4\Delta H_{f(CO_2)} + 3\Delta H_{f(H_2O, l)} - \Delta H_{f(1,3\text{-butadiene})}$

$$\therefore \qquad \Delta H_{f(\text{1,3-butadiene})} = 4\Delta H_{f(\text{CO}_2)} + 3\Delta H_{f(\text{H}_2\text{O},l)} - \Delta H_2$$

Substitution of the two ΔH_f values in equation (1) gives

$$\Delta H = \Delta H_1 - \Delta H_2 - \Delta H_{f(\text{H}_2\text{O},l)} \qquad \Delta H_{f(\text{H}_2,g)} = 0$$

$$= -649 \cdot 8 + 607 \cdot 9 + 68 \cdot 3$$

$$= 26 \cdot 4 \text{ kcal}$$

(ii) Knowing the heat of reaction, the heat of formation of one of the compounds can be determined. This is the way in which the heats of formation of most organic compounds are obtained; the values simply cannot be obtained by a direct route in which the constituent elements combine to form the compound.

EXAMPLE 11.2. *For acetone, the heat of combustion is* $-425 \cdot 7$ *kcal mole^{-1} at 25°C. Calculate the heat of formation of acetone at 25°C.*

$$\text{CH}_3 \cdot \text{CO} \cdot \text{CH}_3(l) + 9/2\,\text{O}_2(g) \rightarrow 3\,\text{CO}_2(g) + 3\,\text{H}_2\text{O}(l)$$
$$\Delta H = -425 \cdot 7 \text{ kcal mole}^{-1}$$

$$\therefore \quad \Delta H = 3\Delta H_{f(\text{CO}_2)} + 3\Delta H_{f(\text{H}_2\text{O},l)} - \Delta H_{f(\text{CH}_3 \cdot \text{CO} \cdot \text{CH}_3)}$$
$$\Delta H_{f(\text{O}_2,g)} = 0$$

$$\therefore \qquad -425 \cdot 7 = -394 \cdot 1 - 368 \cdot 3 - \Delta H_{f(\text{CH}_3 \cdot \text{CO} \cdot \text{CH}_3)}$$

$$\therefore \qquad \Delta H_{f(\text{CH}_3 \cdot \text{CO} \cdot \text{CH}_3)} = -61 \cdot 5 \text{ kcal mole}^{-1}$$

11.5. Fuels: water gas, producer gas, semi-water gas

When steam is blown through a bed of hot coke, *water gas*, essentially a mixture of carbon monoxide and hydrogen, is formed

$$\text{C}(s) + \text{H}_2\text{O}(g) \rightarrow \text{CO}(g) + \text{H}_2(g) \qquad \Delta H = 31 \text{ kcal}$$

This mixture is of medium calorific value† and is a valuable fuel. Because of the endothermic character of the reaction, the continual passage of steam causes the hot coke to cool gradually. If reaction is to continue, an undue drop in the temperature must be prevented. This is achieved by blowing air through the coke, during which time the supply of steam is cut off. The oxygen in the air is first considered to oxidize the carbon to carbon dioxide

$$C(s) + O_2(g) \rightarrow CO_2(g) \qquad \Delta H = -94 \text{ kcal}$$

The carbon dioxide, passing up through the coke, is in turn reduced to the monoxide.

$$C(s) + CO_2(g) \rightarrow 2CO(g) \qquad \Delta H = 41 \text{ kcal}$$

The overall reaction is therefore

$$2C(s) + O_2(g) \rightarrow 2CO(g) \qquad \Delta H = -53 \text{ kcal}$$

Thus during the air blow a mixture of gases consisting mainly of carbon monoxide and nitrogen (from the air) is formed. This mixture is called *producer gas*. Its calorific value is poor because of the diluting effect of the non-combustible nitrogen. Its efficiency as a fuel, however, is improved if it is burned in the hot raw state so that the heat released in its formation (26·5 kcal mole^{-1}) is added to that released during its combustion.

$$CO(g) + \tfrac{1}{2}O_2(g) \rightarrow CO_2(g) \qquad \Delta H = -68 \text{ kcal}$$

The heat which is released during the formation of producer gas causes a rise in bed temperature which, if uncontrolled, can lead to distortion of grate bars and melting of ash. To overcome these difficulties the air supply is cut off at about 1400°C and steam once more blown through. An arrangement of auto-

---

†The efficiency of a fuel is expressed in terms of its *calorific value*, which is the amount of heat liberated when a given amount of fuel is burned completely. Whilst the chemist is inclined to quote a calorific value in terms of kcal g^{-1} (for solid and liquid fuels) and kcal cm^{-3} (for gaseous fuels), the engineer expresses such a quantity in terms of Btu lb^{-1} or Btu ft^{-3}.

matically operated valves ensures the separation of the water gas and producer gas formed alternately in the cyclic process.

Even more convenient in many cases than the cyclic process described above is an arrangement whereby a mixture of steam and air is blown continuously through the hot coke bed in proportions necessary to maintain a constant predetermined temperature. The resulting mixture of gases is referred to as *semi-water gas*. The following example demonstrates how the proper proportion of steam to air may be calculated.

EXAMPLE 11.3. *Determine the proportion by volume of steam to air which would be required to maintain the temperature of the coke at 1000°C. Oxygen content in the air = 20 per cent by volume.*

$$C(s) + H_2O(g) \rightarrow CO(g) + H_2(g) \qquad \Delta H = 31 \text{ kcal}$$

i.e. for every mole of steam decomposed 31 kcal of heat are absorbed.

$$2C(s) + O_2(g) \rightarrow 2CO(g) \qquad \Delta H = -53 \text{ kcal}$$

i.e. for every mole of oxygen decomposed 53 kcal of heat are liberated.

∴ $\frac{31}{53}$ mole of oxygen decomposing will liberate 31 kcal

This is the quantity of oxygen necessary to counteract the cooling which results from reaction between steam and coke.

∴ Moles of air required $= 5 \times \frac{31}{53} = 2 \cdot 925$

∴ Molar proportion of steam = corresponding proportion by
to air   volume
$$= 1 : 2 \cdot 925$$

11.6. Solar heating

Heat from the sun can be used to heat houses. Such dwellings, referred to as solar houses, require the sun's energy to be concentrated and stored so that the energy absorbed during the day may be utilized to heat the house during the night. One such heat-storage unit consists of sealed vessels containing sodium sulphate decahydrate, $Na_2SO_4 \cdot 10H_2O$. At temperatures in excess of $32 \cdot 4°C$ this salt undergoes decomposition according to the equation

$$Na_2SO_4 \cdot 10H_2O \rightarrow Na_2SO_4 + 10H_2O \qquad \Delta H = 19 \cdot 4 \, kcal \qquad (1)$$

The reaction is attended by an enthalpy change of $19 \cdot 4$ kcal. At temperatures below $32 \cdot 4°C$ the reaction proceeds completely in the reverse direction, the consequent enthalpy change being $-19 \cdot 4$ kcal.

$$Na_2SO_4 + 10H_2O \rightarrow Na_2SO_4 \cdot 10H_2O \qquad \Delta H = -19 \cdot 4 \, kcal \qquad (2)$$

Thus during the day the temperature of the sealed containers is maintained at a value in excess of $32 \cdot 4°C$, heat is absorbed from the sun and reaction (1) occurs. At night, the temperature drops below $32 \cdot 4°C$ and the reverse reaction takes place releasing $19 \cdot 4$ kcal for every one mole of $Na_2SO_4 \cdot 10H_2O$ produced.

This system suffers from a number of disadvantages, the most obvious being the dependency on sun exposure. Such a system could hardly be made efficient in the U.K. Even given reasonable exposure to the sun there still remains one serious disadvantage, viz. the problem of heat storage. The reason for this is made clear in the following example.

EXAMPLE 11.4. *Determine the 'calorific value' of* $Na_2SO_4 \cdot 10H_2O$ *when the latter is used as a fuel in the context of house heating.*

$$(1 \, lb = 453 \cdot 6 \, g, 1 \, Btu = 252 \, cal)$$

One mole of Na_2SO_4 produced by the dehydration of one mole of $Na_2SO_4 \cdot 10H_2O$ will release during the night $19 \cdot 4$ kcal or $76 \cdot 98$ Btu. One pound of $Na_2SO_4 \cdot 10H_2O$ will therefore be required to generate 108 Btu, i.e. the 'calorific value' of $Na_2SO_4 \cdot 10H_2O$ is 108 Btu lb^{-1}.

Now this figure is particularly low relative to the calorific values of conventional solid fuels where the values range from 10,000 to 14,000 Btu lb⁻¹. As a result, an extremely large heat-storage unit would be necessary to satisfy the heating requirements of a house of moderate size.

11.7. Rocket propulsion

Thermochemical considerations are important in rocket propulsion technology. A rocket develops its thrust as a result of gas molecules under high pressure being forced to stream through a properly constructed nozzle. This thrust will in turn be dependent on the speed and number of gas molecules produced as a result of reaction between fuel and oxidizer, i.e. oxygen or a source of oxygen. (The combination of rocket fuel and oxidizer is normally referred to as the propellant.) Consequently, an efficient propellant system should have the following characteristics:

  (i) The reaction between fuel and oxidizer should be strongly exothermic.
 (ii) The reaction should be attended by an appreciable increase in the number of gaseous molecules.
(iii) The reaction must be capable of occurring rapidly, much more so than in, for example, a jet engine.
(iv) Both fuel and oxidizer should have a high density in order that storage space be minimized.

The first three features ensure a generation of high-velocity gas molecules under high pressure.

The efficiency of a propellant system is normally expressed in terms of its *specific impulse*, this being the thrust in pounds developed per pound of propellant consumed per second. An approximate criterion of efficiency more rapidly calculated is the enthalpy change per cm³ of propellant.

EXAMPLE 11.5. *On the basis of the above criterion for the efficiency of propellants, show which of the two following systems is the more suitable rocket fuel.*

liquid fuel   oxidizer

(1) $N_2H_5OH(l) + 2H_2O_2(l) \rightarrow N_2(g) + 5H_2O(g)$

Hydroxylamine   Hydrogen
peroxide

(2)   solid fuel   oxidizer

$2LiBH_4(s) + KClO_4(s) \rightarrow Li_2O(s) + B_2O_3(s) +$

Lithium   Potassium $KCl(s) + 4H_2(g)$
borohydride   perchlorate

$\Delta H_{f(N_2H_5OH,l)} = -58$ kcal mole^{-1}, density $= 1\cdot03$ g cm^{-3}

$\Delta H_{f(H_2O_2,l)} = -46$ kcal mole^{-1}, density $= 1\cdot46$ g cm^{-3}

$\Delta H_{f(H_2O,g)} = -57\cdot8$ kcal mole^{-1}

$\Delta H_{f(LiBH_4,s)} = -45$ kcal mole^{-1}, density $= 0\cdot66$ g cm^{-3}

$\Delta H_{f(KClO_4,s)} = -104$ kcal mole^{-1}, density $= 2\cdot52$ g cm^{-3}

$\Delta H_{f(Li_2O,s)} = -142$ kcal mole^{-1}

$\Delta H_{f(B_2O_3,s)} = -302$ kcal mole^{-1}

$\Delta H_{f(KCl,s)} = -104$ kcal mole^{-1}

(1) The enthalpy change ΔH_1 for reaction (1)

$= \Delta H_{f(products)} - \Delta H_{f(reactants)}$

$\therefore \quad \Delta H_1 = 5\Delta H_{f(H_2O,g)} - \Delta H_{f(N_2H_5OH)} - 2\Delta H_{f(H_2O_2)} \quad \Delta H_{f(N_2,g)} = 0$

$= -139$ kcal

This is the quantity of heat liberated when one mole (52 g) of N_2H_5OH reacts with two moles (68 g) of H_2O_2, i.e. 139 kcal are liberated from $(52 + 68)$ g of propellant, i.e. 139 kcal are liberated from $(50\cdot49 + 46\cdot57)$ cm^3 of propellant,

$\therefore \qquad 1\cdot432$ kcal are liberated from 1 cm^3 of propellant

$\therefore \quad$ Enthalpy change per cm^3 of propellant $= -1\cdot432$ kcal

(2) The enthalpy change ΔH_2 for reaction (2) $= \Delta H_{f(Li_2O)} + \Delta H_{(B_2O_3)} + \Delta H_{f(KCl)} - 2\Delta H_{f(LiBH_4)} - \Delta H_{f(KClO_4)} = -354$ kcal, i.e. 354 kcal are liberated from $(43\cdot52 + 138\cdot6)$ g of propellant, i.e. 354 kcal are liberated from $(65\cdot95 + 55\cdot00)$ cm^3 of propellant,

$\therefore \qquad 2\cdot925$ kcal are liberated from 1 cm^3 of propellant

$\therefore \quad$ Enthalpy change per cm^3 of propellant $= -2\cdot925$ kcal

Thus on this basis alone, the second propellant system incorporating a solid fuel is the more efficient. However, combustion of solid fuels is less easy to control than the combustion of a liquid fuel, and this together with other factors must be given due consideration.

11.8. Influence of temperature on equilibrium

In reversible reactions (Chapter 13), the position of equilibrium can be strongly influenced by the temperature. In this connection, a knowledge of both the sign and magnitude of ΔH are of considerable interest. This particular aspect of thermochemistry (important in particular in the field of gaseous fuels) will be discussed fully in Chapter 13.

CHAPTER 12

Chemical Kinetics

CHEMICAL kinetics concerns itself with the speed at which reactions occur. This varies enormously from one reaction to another. For example, the reaction between a strong acid and a strong base is over in an instant, whereas the reaction between hydrogen and oxygen at room temperature proceeds so slowly that one could easily be misled into thinking that no reaction occurs at all. Furthermore, the rate of any one reaction is observed to be influenced by (i) reactant concentration, (ii) temperature, and (iii) the presence of a catalyst.

These observations are explained in terms of the *collision theory*, the basis of which is that *the collision of two or more particles is a prerequisite to chemical reaction*. Using this fundamental assumption, we will consider the sequence of events occurring in the course of a collision between molecules and go on to show (at least qualitatively) why reaction rate should be influenced by the above factors.

12.1. Frequency of collision

The basis of the collision theory is that collision is a prerequisite to reaction. If, however, this were the only requirement then the number of molecules colliding $cm^{-3} sec^{-1}$ would equal the number of molecules reacting $cm^{-3} sec^{-1}$, i.e. the rate of collision would equal the rate of reaction. A comparison of these two quantities, both of which can be calculated, reveals that for most reactions the number of molecules colliding is very much greater than the number of molecules reacting.

Clearly, not all colliding molecules do in fact react upon collision and those which do must differ in some respect from those which do not.

12.2. Energy of collision

We can explain the above inequality by considering in some detail the sequence of events when two molecules collide. Imagine a molecule of hydrogen and one of iodine approaching each other on a collision course. At some point during approach repulsive forces will begin to act due to interaction of the similarly charged electron clouds. During this interaction the potential energy of the system (i.e. its enthalpy) will increase at the expense of its kinetic energy, because work is being done in overcoming the repulsive forces. This increase in potential energy is characterized by, and indeed due to, the gradual weakening of $H-H$ and $I-I$ bonds and to the gradual formation of $H-I$ bonds (Fig. 12.1). A stage is reached when no distinction exists between any of the four bonds. This arrangement is described as an *activated complex* and E, the energy

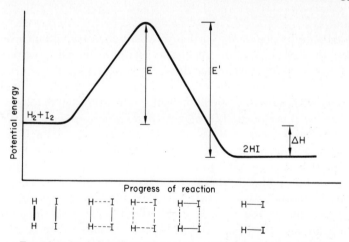

FIG. 12.1. Variation of potential energy with the progress of reaction.

required to produce this state, is defined as the *activation energy*.

The process of bond breaking and bond forming continues as indicated in the above diagram with consequent decrease in potential energy and increase in kinetic energy until the energy of level of the products is reached.

This reaction is reversible (Chapter 13), and if we imagine starting with hydrogen iodide, a similar sequence of events can be traced, producing the same activated complex but a different activation energy E', where $E - E' = \Delta H$.

Hydrogen iodide molecules then can only be formed from hydrogen and iodine molecules via the formation of the activated complex which in turn can only be produced if the energy of the collision is at least equal to the activation energy E. If this is not so then the activated complex will not be formed and the colliding molecules will, having dissipated their kinetic energy, find themselves in the position of possessing potential energy corresponding to the structure

$$\begin{array}{cc} H\text{--}I \\ | \quad | \\ H\text{--}I \end{array}$$

This state is analogous to that of a stretched spring and will, like the spring, tend to change to one of minimum potential energy; i.e. the two molecules will move apart, the kinetic energy increasing at the expense of the potential energy until the latter assumes a value appropriate to that of the reactants. Thus *the energy of the collision determines whether or not reaction will occur.*

Progress is being made but our troubles are not yet over. Why should some collisions be highly energetic while others are not so? You will recall that in a sample of gas the molecules are distributed over a wide range of velocity and hence of kinetic energy (Fig. 4.4). A similar distribution is obtained for collision energies (Fig. 12.2). If E corresponds to the activation energy then the shaded area in Fig. 12.2 represents the

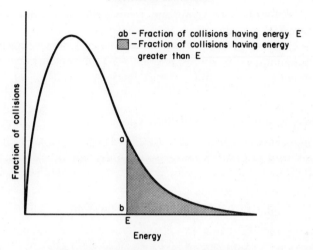

FIG. 12.2. Distribution of collision energies.

fraction of collisions which lead to reaction. For a reaction characterized by a low E value a large proportion of the colliding molecules will suffer decomposition and the rate of the reaction will be high; e.g. the process of neutralization of a strong acid by a strong base requires only the collision between hydrogen and hydroxyl ions, the activation energy being zero and as a result the reaction is so fast as to be almost instantaneous. If, on the other hand, the value of E for a particular reaction is very high only a few collisions will be energetic enough to yield product material and the reaction may then proceed so slowly that no change with time is detectable.

12.3. Geometry of collision

Still one more factor must be considered, viz. an *orientation factor*. Some collisions, though sufficiently energetic, will not result in the formation of an activated complex because the molecules assume during the collision an unsuitable orientation with respect of each other. As would be expected, this effect becomes more significant for a collision occurring between complex molecules.

We may summarize the above by saying that reaction rate is dependent upon

   (i) the frequency of collision,
  (ii) the energy of collision,
 (iii) the geometry of collision,

and any change imposed on a reaction which alters any of these factors will produce a corresponding change in the reaction rate.

12.4.   Influence of concentration

Increasing the concentration of reactants has the effect of increasing reaction rate. This is so because a change in concentration produces a proportional change in the frequency of collision. Whilst the *fraction* of effective collisions remains the same, their *number* changes proportionally.

12.5.   Influence of temperature

The kinetic theory reveals that heat is to be identified with in an increase in collision frequency. For a temperature rise of molecular motion; hence an increase in temperature will result the order of 10 degrees, this increase is calculated to be small and is totally insufficient in itself to account for the very much larger increases in rate which have been observed (sometimes as much as 300 per cent) for such a rise in temperature.

The influence of temperature on the distribution of collision energies, illustrated in Fig. 12.3, follows as a consequence of the effect of temperature on the spread of molecular energies.

From Fig. 12.3 we see that temperature has a pronounced effect on increasing the fraction of collisions capable of generating energy in excess of that required to produce the activated state. Thus, although the frequency of collision is not greatly

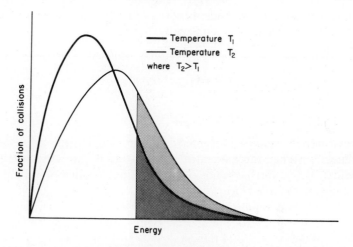

Fig. 12.3. Influence of temperature on the distribution of collision energies.

changed by an increase in temperature, a much greater proportion of the collisions are now sufficiently energetic to lead to product formation.

12.6. Influence of catalyst

A *catalyst* is the name given to a substance which increases a reaction rate without being itself consumed. If reactants and catalyst are in the same phase the catalysis is said to be *homogeneous*.

Consider the lead chamber process for the manufacture of sulphuric acid in which sulphur dioxide is oxidized to sulphur trioxide.

$$2SO_2 + O_2 \rightarrow 2SO_3$$

This reaction, occurring in the gas phase, is extremely slow even at moderate temperatures; it is, however, effectively accelerated by the addition of oxides of nitrogen, principally nitric oxide. The manner in which the nitric oxide involves itself in the reaction is formally indicated in the following equations

$$2NO + O_2 \rightarrow 2NO_2 \tag{1}$$
$$\underline{2SO_2 + 2NO_2 \rightarrow 2SO_3 + 2NO} \tag{2}$$
$$2SO_2 + O_2 \rightarrow 2SO_3 \tag{(1)+(2)}$$

The nitric oxide is consumed in stage 1 and regenerated in stage 2. Hence the amount of nitric oxide is the same at the end as it is at the beginning of the reaction; it is therefore functioning in the role of a catalyst.

In this and similar reactions it is important to note that (i) the catalyst does not initiate reaction, it merely accelerates it, and (ii) the presence of the catalyst causes the reaction to proceed by an alternative route having a quite different rate. These are two important characteristics of all catalysed reactions.

That a reaction can occur in the presence of a catalyst by an alternative route is most significant. In our example, the rate is now determined not by the frequency and effectiveness of collisions between sulphur dioxide and oxygen molecules but by the frequency and effectiveness of collision between (i) nitric oxide and oxygen molecules and (ii) nitrogen dioxide and sulphur dioxide molecules. The concentration of nitric oxide will determine the frequency of collision between reactant molecules (NO and O_2) in stage 1 and reactant molecules (SO_2 and NO_2) in stage 2. Thus the frequency of collision and hence the rate of reaction will be proportional to the concentration of catalyst. This reveals a further characteristic common to all catalysts functioning in a homogeneous capacity.

That the rate of the reaction is appreciably increased where a catalyst is present can only be accounted for by supposing that the alternative route provided is one of lower activation energy (Fig. 12.4). As a result the proportion of effective collisions involving oxygen molecules in stage 1 and of sulphur dioxide molecules in stage 2 will be higher than the correspond-

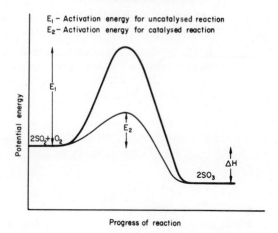

FIG. 12.4. Influence of catalyst on activation energy.

ing proportion of effective collisions between sulphur dioxide and oxygen molecules in absence of catalyst.

The reaction of sulphur dioxide and oxygen is reversible and it should be noted that the activation energy is lowered by an amount $E_1 - E_2$ for *both* forward and reverse reactions. A consequence of this is that a catalyst which accelerates the forward reaction will also accelerate the reverse reaction.

If the reactants and catalyst are not in the same phase then the catalysis is said to be *heterogeneous*. For example, the reaction between carbon dioxide and hydrogen is greatly accelerated by the presence of finely divided platinum, the latter functioning as a catalyst.

$$CO_2 + H_2 \xrightarrow{Pt} H_2O + CO$$

This acceleration is found to be dependent on the amount of catalyst surface available to the reactant materials, the rate increasing with increase in surface area. The mechanism by which this and similar reactions are catalysed heterogeneously cannot be described in quite such precise terms as is possible for reactions catalysed homogeneously. The sequence of events for the catalysed reaction between carbon dioxide and hydrogen are considered to involve three steps.

(a) *The adsorption of reactant molecules*
on the surface

The process of adsorption† can be regarded somewhat loosely as reaction between the reactant molecules and the surface atoms of catalyst.

$$\left.\begin{array}{c} CO_2 + Pt \rightarrow CO_2(Pt) \\ H_2 + Pt \rightarrow H_2(Pt) \end{array}\right\} \text{Adsorbed reactants.}$$

As a result of interaction between the surface atoms of catalyst and reactant molecules there may be a weakening of particular bonds in the adsorbed molecules thus making them more susceptible to reaction, i.e. they have become activated in the process of adsorption.

(b) *Surface reaction*

Reaction occurs on the surface between the now activated molecules provided they occupy adjacent positions on the surface

$$\underbrace{CO_2(Pt) + H_2(Pt)}_{\text{adsorbed reactants}} \rightarrow \underbrace{H_2O(Pt) + CO(Pt)}_{\text{adsorbed products}}$$

(c) *Desorption of products*

Product molecules break free from the surface, i.e. are desorbed, and continue their existence in the gaseous state.

$$H_2O(Pt) \rightarrow H_2O + Pt$$
$$CO(Pt) \rightarrow CO + Pt$$

†Adsorption is the term used to describe the adhesion of molecules on the *surface* of a solid. Any increase in surface area will lead to an increase in adsorption.

Essentially the presence of a catalyst has provided (as before) an alternative route having an activation energy lower than that of the non-catalysed reaction. This decrease in activation energy accounts for the increase in reaction rate.

CHAPTER 13

Chemical Equilibria

IT IS an experimentally observed fact that for many reactions the interacting substances are not completely consumed. Thus, if hydrogen and iodine are heated in a closed vessel, they will react only partially to produce hydrogen iodide.

$$H_2(g) + I_2(g) \rightarrow 2HI(g)$$

An essential requirement for this and indeed any reaction to occur is that the reacting molecules should collide. Now only a fraction of the collisions will result in the formation of hydrogen iodide molecules and as their number increases, the probability of a collision occurring between two hydrogen iodide molecules likewise increases. A fraction of these collisions will lead to the formation of hydrogen and iodine.

$$2HI(g) \rightarrow H_2(g) + I_2(g)$$

A stage is finally reached when no further change in overall composition with time is observed. The system is then said to be in a state of *dynamic equilibrium*. As a result of molecular collision the reaction continues but in such a way that for every molecule of hydrogen which reacts through collision with one molecule of iodine, two molecules of hydrogen iodide decompose elsewhere in the system. There is therefore an equality between the rate at which hydrogen iodide is formed and the rate at which it is decomposed. Such a reaction is sensibly called *reversible* and is represented in the chemical equation by the symbol $\rightleftharpoons$ as in

$$H_2(g) + I_2(g) \rightleftharpoons 2HI(g)$$

It is possible that all reactions are reversible – the problem is to select the conditions which serve to reveal the reversible character.

13.1. Equilibrium constants

Consider the general case of an equilibrium involving different numbers of reactant and product molecules.

$$a\text{A} + b\text{B} \cdots \rightleftharpoons c\text{C} + d\text{D} \cdots$$

It is found experimentally *that for any reversible reaction under equilibrium conditions, the ratio of the product of product concentrations to the product of reactant concentrations (each concentration being raised to a power equal to the coefficients in the balanced equation) is a constant characteristic of the reaction and dependent only on the temperature.* The above statement, referred to as the *Law of Chemical Equilibrium*, is the foundation upon which an understanding of chemical equilibria is based.

The constant referred to in the above law is called the *equilibrium constant* and is related to the concentrations of reactants and products by the expression

$$K = \frac{[\text{C}]^c [\text{D}]^d \cdots}{[\text{A}]^a [\text{B}]^b \cdots}$$

where [ ] represents the equilibrium concentration (mole 1^{-1}) of whatever component is included within the bracket.

The fact that the product concentrations appear in the numerator while reactant concentrations form the denominator is a purely arbitrary choice – obviously we must be consistent and this convention is the one universally accepted.

EXAMPLE 13.1. *Write down the equilibrium constant expression for each of the following reactions:*

$$H_2(g) + I_2(g) \rightleftharpoons 2HI(g)$$
$$2HI(g) \rightleftharpoons H_2(g) + I_2(g)$$

$$2SO_2(g) + O_2(g) \rightleftharpoons 2SO_3(g)$$
$$4NO(g) + 6H_2O(g) \rightleftharpoons 4NH_3(g) + 5O_2(g)$$

For the reaction

$$H_2(g) + I_2(g) \rightleftharpoons 2HI(g)$$
$$K = \frac{[HI]^2}{[H_2][I_2]}$$

The reaction written in this way implies that the starting materials are $H_2(g)$ and $I_2(g)$. If the starting material was instead $HI(g)$, the equation would be more reasonably written as

$$2HI(g) \rightleftharpoons H_2(g) + I_2(g)$$

for which

$$K' = \frac{[H_2][I_2]}{[HI]^2} = \frac{1}{K}$$

For the reaction

$$2SO_2(g) + O_2(g) \rightleftharpoons 2SO_3(g)$$
$$K = \frac{[SO_3]^2}{[SO_2]^2[O_2]}$$

For the reaction

$$4NO(g) + 6H_2O(g) \rightleftharpoons 4NH_3(g) + 5O_2(g)$$
$$K = \frac{[NH_3]^4[O_2]^5}{[NO]^4[H_2O]^6}$$

The extent to which a reaction proceeds from left to right can be deduced from the magnitude of the equilibrium constant; if large, then the concentrations of at least one of the products must be high relative to those of the reactants, indicating that the position of equilibrium lies well over to the right. Conversely a small equilibrium constant implies that the reaction proceeds from left to right to only a small extent.

13.2. The equilibrium constant K_p

If a reaction occurs in the gas phase, it is convenient to express the concentrations of components present by their respective partial pressures. That this is justifiable can be easily demonstrated. For a gas A in a mixture

$$P_A = \frac{n_A RT}{V}$$

where

$$P_A = \text{partial pressure of A}$$

and

$$\frac{n_A}{V} = \text{molar concentration of A} = [A].$$

$\therefore$ At constant temperature $P_A \propto [A]$

The equilibrium constant for the reaction

$$aA + bB \cdots \rightleftharpoons cC + dD \cdots$$

is then written as

$$K_p = \frac{(P_C)^c (P_D)^d \cdots}{(P_A)^a (P_B)^b \cdots}$$

where P_A, P_B, P_C and P_D are the partial pressures of A, B, C and D respectively. It is advisable at all times to express partial pressures in atmospheres.

13.3. Relationship between K_p and K

Consider once more the general equation

$$aA + bB \cdots \rightleftharpoons cC + dD \cdots$$

in which, for a reaction occurring in the gas phase,

$$P_A = \frac{n_A RT}{V} = [A]RT$$

Similarly $P_B = [B]RT, P_C = [C]RT, P_D = [D]RT$

$$K_p = \frac{(P_C)^c (P_D)^d \cdots}{(P_A)^a (P_B)^b \cdots} = \frac{[C]^c [D]^d (RT)^{(c+d-a-b)}}{[A]^a [B]^b} = K(RT)^{\Delta n}$$

where $\Delta n = (c + d - a - b) =$ the change in the number of gaseous moles.

EXAMPLE 13.2. *For the decomposition of nitrogen tetroxide to nitrogen dioxide at* 318°K, *according to the equation*

$$N_2O_4(g) \rightleftharpoons 2NO_2(g)$$

$K_P = 0.6533$ *when partial pressures are expressed in atmospheres. What is the value of K at this temperature?*

$$K_p = K(RT)^{\Delta n} = 0.6533$$

$\Delta n = +1$, and $R = 0.082$ l. atm deg^{-1} mole^{-1}

$$\therefore \qquad K = \frac{0.6533}{0.082 \times 318} = 0.02505$$

13.4.   Heterogeneous equilibria

A heterogeneous equilibrium is one in which two or more phases are present. For example, the reaction between steam and red-hot coke involves both gaseous and solid phases

$$C(s) + H_2O(g) \rightleftharpoons CO(g) + H_2(g)$$

whence

$$K' = \frac{[CO][H_2]}{[C][H_2O]}$$

The concentration of a solid is numerically identical to its density when the latter is expressed in units of mole l^{-1}. Clearly the density of a solid is not dependent on the amount of solid present, and we can therefore combine this constant concentration with the original equilibrium constant,

i.e.

$$K'[C] = \frac{[CO][H_2]}{[H_2O]} = K$$

Thus, in any heterogeneous equilibrium, concentrations of pure solids do not appear in the equilibrium constant expression; nor, by a similar argument, do the concentrations of pure liquids.

13.5.   Influence of concentration on the equilibrium position

Consider once more the reaction

$$C(s) + H_2O(g) \rightleftharpoons CO(g) + H_2(g)$$

for which

$$K = \frac{[CO][H_2]}{[H_2O]}$$

Suppose that to an equilibrium mixture more steam is added; if the temperature is kept constant, the value of the equilibrium constant must remain the same and it can only do so if some of the additional steam reacts with carbon to produce the appropriate amounts of carbon monoxide and hydrogen, thereby displacing the equilibrium position to the right. *Generally, when the concentration of one of the components making up an equilibrium system is increased the reaction promoted is that which tends to consume some of the added substance.*

The same result could be obtained by removing the products from the sphere of action. Their continued removal would cause more steam to react with carbon in order that the equilibrium constant should maintain its value, and in this way the reaction could be forced to completion.

EXAMPLE 13.3. *For the reaction*

$$H_2(g) + I_2(g) \rightleftharpoons 2HI(g)$$

$$K = 50 \cdot 25 \text{ at } 717°K$$

(i) *Calculate the number of moles of* HI *formed when two moles of* H_2 *and one mole of* I_2 *react at* 717°K *under a condition of constant volume.*

(ii) *Compare this result with that obtained when a further one mole of* H_2 *is introduced into the system.*

(i) Let $x =$ moles of H_2 undergoing decomposition.
The equilibrium concentrations are therefore as follows:

$$[H_2] = \frac{2-x}{V}, [I_2] = \frac{1-x}{V}, [HI] = \frac{2x}{V}$$

∴ $$K = \frac{[HI]^2}{[H_2][I_2]} = \frac{4x^2}{(2-x)(1-x)} = 50 \cdot 25$$

∴ $$46 \cdot 25x^2 - 150 \cdot 75x + 100 \cdot 5 = 0$$

This is a quadratic equation of the form

$$ax^2 + bx + c = 0$$

the solution of which is

$$x = \frac{-b \pm \sqrt{(b^2 - 4ac)}}{2a}$$

Following this procedure

$$x - 2 \cdot 327 \quad \text{or} \quad 0 \cdot 9335$$

As there were only two moles of H_2 initially present, a value of $x = 2 \cdot 327$ is clearly ridiculous. Hence the number of HI moles present at equilibrium $= 2x = 1 \cdot 8670$

(ii) Let y = moles of H_2 undergoing decomposition.

The equilibrium concentrations are therefore as follows:

$$[H_2] = \frac{3-y}{V}, [I_2] = \frac{1-y}{V}, [HI] = \frac{2y}{V}$$

$$\therefore \qquad K = 50 \cdot 25 = \frac{4y^2}{(3-y)(1-y)}$$

$$\therefore \qquad y = 3 \cdot 383 \text{ or } 0 \cdot 9645$$

The second value is the only permissible one, and so the number of HI moles present at equilibrium is $1 \cdot 9390$. It is evident from a comparison of the two results that increasing the concentration of HI forces the equilibrium to the right.

13.6. Influence of pressure on the equilibrium position

We have already seen that for the general equation

$$aA + bB \cdots \rightleftharpoons cC + dD \cdots$$

$$K_P = \frac{(P_C)^c (P_D)^d \cdots}{(P_A)^a (P_B)^b \cdots}$$

Now $P_A = x_A P, P_B = x_B P, P_C = x_C P, P_D = x_D P$ where P is the total pressure and x_A, x_B, x_C, and x_D are the mole fractions of A, B, C and D respectively.

$$\therefore \qquad K_P = \frac{x_C{}^c \cdot x_D{}^d}{x_A{}^a \cdot x_B{}^b} \cdot P^{\Delta n} = A \cdot P^{\Delta n} \text{ where } A = \frac{x_C{}^c \cdot x_D{}^d}{x_A{}^a \cdot x_B{}^b}$$

This equation reveals the influence that pressure has on the position of equilibrium.

Consider a reaction characterized by a positive value of Δn. If the pressure acting on the system is increased, the A term must decrease if the value of the equilibrium constant is not to change. This decrease can only be achieved if C and D react to some extent (thereby reducing the numerator) to produce A and B (thereby lowering the denominator). Thus for such a reaction, an increase in pressure will favour the formation of reactant materials, the equilibrium position being displaced to the left. A similar argument leads to the conclusion that for a reaction having a negative value of Δn, an increase in pressure causes a shift of the equilibrium position to the right. For a reaction for which Δn is zero, a change in pressure has no effect on the equilibrium position, and this reflects itself in the equality between the terms K and K_P.

Summarizing, we can say that *an increase in pressure favours the formation of those products formed with a decrease in the number of gaseous molecules; a decrease in pressure favours the formation of those products formed with an increase in the number of gaseous molecules; a change of pressure has no effect on a system for which there is no change in the number of gaseous molecules.*

Dissociation equilibria provide a good illustration of a group of reactions for which Δn is not equal to zero. The breakdown, i.e. dissociation, of nitrogen tetroxide into nitrogen dioxide exemplifies this type of reaction.

$$N_2O_4(g) \rightleftharpoons 2NO_2(g)$$

For such a reaction, the fraction of one mole undergoing complete decomposition is referred to as the *degree of dissociation*, normally represented by α.

Thus, if a moles of N_2O_4 are initially present, then:

$a\alpha$ = moles of N_2O_4 undergoing complete decomposition,
$a - a\alpha = a(1-\alpha)$ = moles of N_2O_4 present at equilibrium,
and $2a\alpha$ = moles of NO_2 present at equilibrium.

$\therefore$ $\quad (a - a\alpha) + 2a\alpha = a(1+\alpha)$ = total moles present at equilibrium.

$\therefore$ $$P_{N_2O_4} = \frac{n_{N_2O_4}}{n_T} \cdot P = \frac{a(1-\alpha)}{a(1+\alpha)} \cdot P = \frac{1-\alpha}{1+\alpha} \cdot P$$

and $$P_{NO_2} = \frac{n_{NO_2}}{n_T} \cdot P = \frac{2a\alpha}{a(1+\alpha)} \cdot P = \frac{2\alpha}{1+\alpha} \cdot P$$

$\therefore$ $$K_P = \frac{(P_{NO_2})^2}{(P_{N_2O_4})} = \frac{4\alpha^2}{1-\alpha^2} \cdot P$$

EXAMPLE 13.4. *At 318°K and 1 atm, nitrogen tetroxide is 37·8 per cent dissociated. To what extent would it be dissociated at 318°K and 2 atm?*

The degree of dissociation is given here as a percentage. This means that for every 100 moles of N_2O_4, 37·8 moles undergo dissociation. Hence the fraction of one mole dissociating (α) is 0·378.

$$K_P = \frac{4\alpha^2}{1-\alpha^2} \cdot P = \frac{4 \times 0·378^2 \times 1}{1 - 0·378^2} = 0·6533$$

The new condition of pressure does not alter the value of K_p. Hence

$$K_P = 0·6533 = \frac{8\alpha^2}{1-\alpha^2}$$

and

$$\alpha = 0·2742$$

An increase of pressure has, in agreement with our previous observation, reduced the value of α as a result of the equilibrium position being displaced to the left.

13.7. Influence of temperature on the equilibrium position

Changing the concentration of materials involved in an equilibrium or altering the pressure acting on the system can cause a displacement of the equilibrium position. There is, however, no consequent change in the value of the equilibrium constant unless the temperature changes.

By an argument beyond the scope of this book, it can be shown that the dependence of K_P on temperature is given by the equation

$$\frac{d \ln K_P}{dT} = \frac{\Delta H}{RT^2}$$

where R is the familiar gas constant and ΔH is the enthalpy change, i.e. the heat evolved (or absorbed) under a condition of constant pressure when the reaction proceeds *completely* from left to right.

The interesting feature of the above equation is that it indicates the influence of the sign and magnitude of the enthalpy term in determining the manner in which the equilibrium constant varies with temperature. Thus for an exothermic reaction, for which ΔH is negative, $d \ln K_P$ is negative when dT is positive. Hence K_P will decrease with increase in temperature and the equilibrium will be displaced in the direction which *absorbs* heat. If, on the other hand, ΔH is positive (endothermic reactions) then $d \ln K_P/dT$ will be positive, indicating that K_P will increase with increase in temperature, the equilibrium position again being displaced in the direction which absorbs heat. Conversely, a decrease in temperature (for which dT is negative) can be shown to displace the position of equilibrium in the direction which liberates heat. In short, *an increase in temperature favours the formation of those products formed with an absorption of heat, whilst a decrease in temperature favours the formation of those products formed with a liberation of heat.*

If it is desired to know values of K_P for a variety of tempera-tures, then the equation may be integrated between the limits of temperature T_1 and T_2. Thus, if $(K_P)_1$ and $(K_P)_2$ are the equilibrium constants corresponding to temperatures T_1 and T_2, then

$$\int_{(K_P)_1}^{(K_P)_2} d \ln K_P = \int_{T_1}^{T_2} \frac{\Delta H}{RT^2} \cdot dT$$

If we assume the temperature range to be not too large, ΔH can be regarded as a constant.

$$\therefore \qquad \ln \frac{(K_P)_2}{(K_P)_1} = \frac{\Delta H}{R} \int_{T_1}^{T_2} \frac{dT}{T^2} = \frac{\Delta H}{R} \cdot \left(\frac{T_2 - T_1}{T_1 T_2} \right)$$

$$\therefore \qquad \log \frac{(K_P)_2}{(K_P)_1} = \frac{\Delta H}{2 \cdot 303 R} \left(\frac{T_2 - T_1}{T_1 T_2} \right) \qquad \ln x = 2 \cdot 303 \log x$$

EXAMPLE 13.5. *For the reaction*

$$N_2(g) + O_2(g) \rightleftharpoons 2NO(g)$$

$K_p = 2 \cdot 31 \times 10^{-4}$ *at* 1900°K *and* $\Delta H = 21 \cdot 60$ *kcal mole*$^{-1}$. *What is the value of* K_P *at* 2300°K?

The value of ΔH given represents the heat absorbed (under a condition of constant pressure) when *one* mole of NO is produced from its elements; hence, when two moles are formed 43·20 kcal will be absorbed.

$$\therefore \qquad \log \frac{K_{2300}}{K_{1900}} = \frac{43,200}{2 \cdot 303 \times 1 \cdot 987} \left(\frac{2300 - 1900}{2300 \times 1900} \right) = 0 \cdot 8642$$

$$\therefore \qquad \frac{K_{2300}}{K_{1900}} = 7 \cdot 314 \text{ and } K_{2300} = 16 \cdot 9 \times 10^{-4}$$

This result is consistent with our previous observations in that the equilibrium position has been displaced in such a way as to increase the concentrations of those products formed with the absorption of heat.

13.8. Influence of catalyst on equilibrium position

In Chapter 12 we noted that in a reversible reaction, a catalyst which accelerates the forward reaction will also accelerate the reverse reaction. In fact, the two opposing rates change proportionally. The net effect is to hasten the establishment of equilibrium without altering the equilibrium concentrations or the equilibrium constant.

13.9. Industrial manufacture of ammonia

Ammonia is produced industrially by the reaction between nitrogen and hydrogen.

$$N_2 + 3H_2 \rightleftharpoons 2NH_3$$

This reversible reaction is characterized in the forward direction by a reduction in the number of gaseous moles ($\Delta n = -2$) and by the liberation of heat ($\Delta H = -22 \, \text{kcal}$). Consider the consequences of these characteristics with a view to predicting the conditions most likely to result in a maximum yield of ammonia.

(a) *Equilibrium considerations*

The equilibrium constant for the ammonia synthesis reaction is given by

$$K_P = \frac{(P_{NH_3})^2}{(P_{N_2})(P_{H_2})^3} = \frac{(x_{NH_3})^2}{(x_{N_2})(x_{H_2})^3} \cdot (P)^{-2} = A \cdot P^{-2}$$

The above relationship indicates the influence of pressure on the position of equilibrium. In this case, an increase in pressure must result in an attendant increase in the A term if the equilibrium constant is to maintain its value appropriate to a particular temperature. *High pressure will therefore favour the formation of ammonia.*

The influence of temperature on the equilibrium position can be deduced from the equation

$$\frac{d \ln K_P}{dT} = \frac{\Delta H}{RT^2}$$

When ΔH is negative (exothermic reaction) a decrease in temperature must lead to an increase in the value of K_P. *Thus the use of low temperatures will favour the formation of ammonia.*

Our predictions regarding the influence of pressure and temperature on the position of equilibrium are confirmed in the following graph (Fig. 13.1) in which experimental values of the equilibrium percentage of ammonia in a 3:1 hydrogen–nitrogen gas mixture are plotted against the pressure at a variety of temperatures.

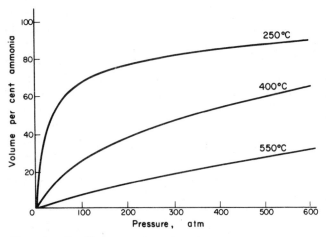

FIG. 13.1. Equilibrium concentration of ammonia in the reaction between nitrogen and hydrogen.

(b) *Rate considerations*

Equilibrium considerations indicate *how far* we can expect the reaction to proceed to the right under certain conditions of pressure and temperature. Such information will not, however, tell us *how rapidly* the equilibrium will establish itself. Thus a decrease in temperature, although increasing the equilibrium percentage of ammonia, decreases the rate at which nitrogen and hydrogen react, so much so that at temperatures below 300°C the reaction would not be economically feasible. Nor would the situation be improved by using very high temperatures for although the rate would markedly increase (as a result of an increase in the proportion of effective collisions) the equilibrium percentage of ammonia would be decreased.

Circumstances such as this suggest the use of a catalyst, in this case an iron oxide. In this way, the activation energy for the reaction is decreased and the temperature which optimizes the rate and yield is in the region of 500°C. The pressures used vary from 100 to 1000 atm.

At present, all plants manufacturing ammonia use modifications of the original Haber process introduced at the turn of this century. The sources of nitrogen and hydrogen are many and varied. Two common sources are *producer gas* $(CO + H_2)$ and *water gas* $(CO + H_2)$.

To obtain gas containing the proper proportions of nitrogen and hydrogen, a gas 'producer' alternately manufacturing producer and water gas is used (Section 11.5). These two gases are mixed in such proportions as to yield after suitable treatment nitrogen and hydrogen in the ratio of 1 : 3.

13.10. Producer gas

Producer gas (Section 11.5) is obtained when air is passed through a bed of white-hot coke, the sequence of reactions being

(1) $C + O_2 \rightarrow CO_2$ $\quad \Delta H = -94 \text{ kcal}$

(2) $C + CO_2 \rightleftharpoons 2CO$ $\quad \Delta H = 41 \text{ kcal}$

Reaction (1) proceeds to virtual completion; reaction (2), however, is reversible and an equilibrium is set up in the gas producer between the carbon, carbon monoxide and carbon dioxide. Irrespective of whether the producer gas is to be used as a fuel or as a source of nitrogen, it is desirable to have as little carbon dioxide present in the final mixture of gases as possible. We wish, in other words, to select conditions which will generate as high a ratio of carbon monoxide to carbon dioxide as possible. In view of the fact that the carbon monoxide is formed with the absorption of heat, its formation will therefore be favoured by employing high temperatures. Such a prediction is confirmed by the experimental data (Fig. 13.2). From these results it is clear that the temperature of the producer should not be allowed to fall below 1000°C if the proportion of carbon dioxide is to be kept to a minimum. (The condition of high temperature has, of course, the added advantage of increasing reaction rate.) If this temperature is maintained, effectively all the carbon dioxide produced in reaction (1) is consumed in reaction (2) and the overall heat change is moderately exothermic.

$$
\begin{array}{lll}
C + O_2 & \rightarrow CO_2 & \Delta H = -94 \text{ kcal} \\
C + CO_2 & \rightarrow 2CO & \Delta H = 41 \text{ kcal} \\
\hline
2C + O_2 & \rightarrow 2CO & \Delta H = -53 \text{ kcal}
\end{array}
$$

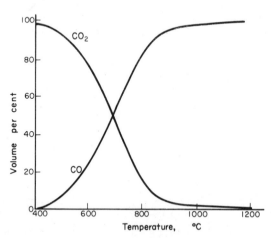

Fig. 13.2. Equilibrium concentration of gases in the reaction between carbon and oxygen.

13.11. Water gas

The passage of steam through a bed of hot coke results in the production of water gas (Section 11.5). The reaction between steam and coke is reversible.

(1) $$C + H_2O \rightleftharpoons CO + H_2 \qquad \Delta H = 31 \text{ kcal}$$

Some carbon dioxide is always present due to the reaction

(2) $$CO + H_2O \rightleftharpoons CO_2 + H_2 \qquad \Delta H = -10 \text{ kcal}$$

As with producer gas, the carbon dioxide content should be as low as possible. This condition can be realized by maintaining a high temperature which in addition to producing high reaction rate will favour the formation of products in the endothermic reaction (1) and will suppress the formation of products in the exothermic reaction (2), the equilibrium position being shifted well over to the left.

The manner in which the composition of the system changes with temperature (Fig. 13.3) clearly confirms the need to maintain a temperature of at least 1000°C if the carbon dioxide content of the mixture is to be minimal.

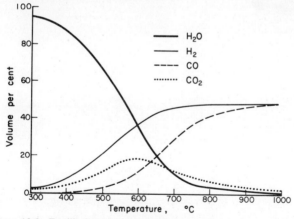

FIG. 13.3. Equilibrium concentration of gases in the reaction between steam and carbon. (Munro, L. A., *Chemistry in Engineering*, Prentice-Hall, New Jersey, 1964).

13.12. Catalytic decomposition of water gas

In the production of water gas, a condition of temperature was chosen so as to minimize the extent to which the carbon monoxide was decomposed by steam

$$CO + H_2O \rightleftharpoons CO_2 + H_2 \qquad \Delta H = -10 \text{ kcal}$$

If it is desired to obtain hydrogen by this reaction (using water gas as a source of carbon monoxide) then conditions have to be chosen which facilitate the decomposition of carbon monoxide. Thus, while the yield of hydrogen will be clearly favoured by a low temperature (the reaction being exothermic) it will be unaffected by whatever pressure is used, there being no change in the number of gaseous moles ($\Delta n = 0$).

The way in which the proportion of gases present changes with temperature (Fig. 13.4) shows that at low temperatures very little carbon monoxide remains and for every one volume of carbon dioxide present there are approximately two volumes of hydrogen (one being formed by reaction, the other being the amount initially present in the water gas). At a temp-

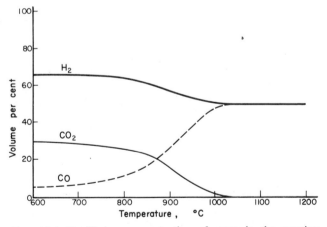

FIG. 13.4. Equilibrium concentration of gases in the reaction between carbon monoxide and steam.

erature of 1200°C, very little carbon dioxide is present and the carbon monoxide and hydrogen are present in approximately the same volume proportion; i.e. the original water gas experiences very little decomposition at this temperature.

Although from a consideration of equilibrium factors a low temperature is ideal, such a condition is unsuitable from the point of view of reaction rate. Once more, a compromise must be made. In this case, a temperature of 450–500°C gives a reasonable rate with the help of a catalyst (ferric oxide and chromium oxide) together with a good equilibrium yield of hydrogen, which indeed can still further be improved by increasing the proportion of steam. This results, in accordance with the law of chemical equilibrium, in a displacement of the equilibrium position from left to right.

CHAPTER 14

Acids and Bases

14.1. Dissociation constants

Although some characteristics of acids and bases (e.g. bitter taste, corrosive action) are well known, these properties cannot be used to define the terms 'acid' and 'base'. Perhaps the most widely used definition of these terms is that an acid is a substance which gives rise to hydrogen ions in solution while a base produces hydroxyl ions:

$$HNO_3 \rightarrow H^+ + NO_3^-$$
$$KOH \rightarrow K^+ + OH^-$$

Both the hydrogen and hydroxyl ions will be solvated, i.e. will be fairly strongly associated with one or more solvent molecules. In particular, the very small hydrogen ion H^+ is not very stable, and in aqueous solution will form a co-ordinate linkage with a water molecule to form a *hydronium ion* H_3O^+.

$$H^+ + O\begin{array}{c} H \\ \diagup \\ \diagdown \\ H \end{array} \rightarrow [H-\overset{\overset{\textstyle H}{|}}{O}-H]^+$$

On this basis the characteristic properties of acids are in fact the properties of the hydronium ion, and since the common acids are covalent molecules in the pure state, they do not then behave as acids. Pure anhydrous hydrogen chloride, for example, will have no effect on litmus.

Another point to note is that the formula of any compound gives little indication of its properties. If we consider the three materials phenol (C_6H_5OH), methanol (CH_3OH) and sodium hydroxide (NaOH), we find that although each possesses an $-OH$ group, phenol behaves as an acid, sodium hydroxide as a base, and methanol as neither.

Since acidic properties have been defined as the properties of the hydrogen (or hydronium) ion, it follows that the greater the concentration of hydrogen ions in solution, the more pronounced will be the acidity. If a molecule dissociates in solution to form hydrogen ions, the extent of dissociation will be governed by the equilibrium constant for the reaction:

$$HA \rightleftharpoons H^+ + A^-$$

$$K_a = \frac{[H^+][A^-]}{[HA]}$$

This equilibrium constant is called the *dissociation constant* of the acid.

In aqueous solution at 25°C the value of K_a for HCl is of the order of $1 \cdot 0 \times 10^7$, while K_a for acetic acid is $1 \cdot 8 \times 10^{-5}$. In aqueous solution, therefore, HCl is almost completely dissociated into ions and is said to be a *strong acid*, while acetic acid is only slightly dissociated and is described as a *weak acid*.

Similarly the strength of a base may be classified by the value of the equilibrium constant for the reaction

$$MOH \rightleftharpoons M^+ + OH^-$$

$$K_b = \frac{[M^+][OH^-]}{[MOH]}$$

Again we find *strong bases* (e.g. sodium hydroxide) with very large dissociation constants, and *weak bases* (e.g. ammonium hydroxide) which are only weakly dissociated.

14.2. Proton transfer theory

Although the definition of acids and bases outlined above is a very useful one, particularly for aqueous solutions, it has certain limitations. One of the most serious is that bases are limited to compounds which can give rise to hydroxyl ions. Further-more, many compounds which cannot be classified as acids or bases on our simple definition exhibit acidic or basic properties in for example catalytic reactions, or in solvents other than water.

For these reasons the acid-base concept was broadened by Brønsted and Lowry, independently in 1923. They defined an acid as any substance which can give up a proton, while a base is any substance which can accept a proton. In other words, an acid is a *proton donor*, and a base is a *proton acceptor*. On this basis, the relationship between an acid and a base is given by the equation:

$$A \rightleftharpoons B + H^+$$
$$\text{Acid} \quad \text{Base} \quad \text{Proton}$$

Here the acid A and base B which differ from one another by a proton, are described as a *conjugate pair*, i.e. the base B is said to be the conjugate base of the acid A.

We have already seen that a proton does not exist as such in solution, but combines with a water molecule to form the hydro-nium ion. In this case the water molecule has acted as a proton acceptor, that is, as a base. Whereas on the simple theory we would write

$$HNO_3 \rightleftharpoons H^+ + NO_3^-$$
$$CH_3CO_2H \rightleftharpoons H^+ + CH_3CO_2^-$$
$$HSO_4^- \rightleftharpoons H^+ + SO_4^{2-}$$

On the proton transfer picture these equations would be chang-ed to

$$HNO_3 + H_2O \rightleftharpoons H_3O^+ + NO_3^-$$
$$\text{Acid}_1 \quad \text{Base}_2 \quad \text{Acid}_2 \quad \text{Base}_1$$

$$CH_3CO_2H + H_2O \rightleftharpoons H_3O^+ + CH_3CO_2^-$$
$$\text{Acid}_1 \quad \text{Base}_2 \quad \text{Acid}_2 \quad \text{Base}_1$$

$$HSO_4^- + H_2O \rightleftharpoons H_3O^+ + SO_4^{2-}$$
$$\text{Acid}_1 \quad \text{Base}_2 \quad \text{Acid}_2 \quad \text{Base}_1$$

In each of these reactions water acts as a base. Furthermore, the nitrate, sulphate and acetate ions are classified as bases, because each is capable of accepting a proton to form its conjugate acid.

The proton transfer theory interprets the term 'base' much more broadly than did the simple dissociation theory, which limited the definition to solutions of OH^- ions. From the few examples quoted above it can be seen that the OH^- ion is only one of a large number of bases, and while compounds such as NaOH are still described as bases, it is the OH^- ion which is the active constituent.

We have described how water can act as a proton acceptor, i.e. as a base:

$$\underset{\text{Base}}{H_2O} + H^+ \rightleftharpoons \underset{\text{Acid}}{H_3O^+}$$

but it can also act as a proton donor, i.e. an acid

$$\underset{\text{Acid}}{H_2O} \rightleftharpoons H^+ + \underset{\text{Base}}{OH^-}$$

These two equations may be combined to give

$$\underset{\text{Acid}_1}{H_2O} + \underset{\text{Base}_2}{H_2O} \rightleftharpoons \underset{\text{Acid}_2}{H_3O^+} + \underset{\text{Base}_1}{OH^-}$$

In other words, water undergoes a self-ionization process and the acidity or basicity of any aqueous solution can be regarded as the result of disturbing this equilibrium by the addition of some other material. The well-known neutralization reaction of an acid with a base, once defined by the equation

$$\text{Acid} + \text{Base} \rightarrow \text{Salt} + \text{Water}$$

is simply the reaction between hydronium and hydroxyl ions, i.e. the reverse reaction written above.

It must be appreciated that water is by no means unique in its ability to undergo self-ionization, and that other solvents can act in this same way. Ammonia, for example, can act either as an acid, or as a base:

$$\underset{\text{Acid}}{NH_3} \rightleftharpoons \underset{\text{Base}}{NH_2^-} + H^+$$

$$\underset{\text{Base}}{NH_3} + H^+ \rightleftharpoons \underset{\text{Acid}}{NH_4^+}$$

14.3. Lewis acids and bases

The acid-base concept can be broadened even further if the definition given by G. N. Lewis is adopted. He defined an acid as any substance which is capable of *accepting* an electron pair, and a base as any substance capable of *donating* an electron pair. Thus in the reaction

$$H^+ +: \overset{\overset{\displaystyle H}{\cdot|\cdot}}{\underset{\underset{\displaystyle H}{\cdot|\cdot}}{N}} \div H \rightleftharpoons \left[H \div \overset{\overset{\displaystyle H}{\cdot|\cdot}}{\underset{\underset{\displaystyle H}{\cdot|\cdot}}{N}} \div H \right]^+$$

H^+ is the acid and NH_3 the base because of the formation of a co-ordinate bond. Now, however, substances containing no hydrogen at all may also be classified as acids, as in the reaction

$$\overset{\overset{\displaystyle F}{\cdot|\cdot}}{\underset{\underset{\displaystyle F}{\cdot|\cdot}}{F \div B}} +: \overset{\overset{\displaystyle H}{\cdot|\cdot}}{\underset{\underset{\displaystyle H}{\cdot|\cdot}}{N}} \div H \rightleftharpoons \overset{\overset{\displaystyle F \quad H}{\cdot|\cdot \quad \cdot|\cdot}}{\underset{\underset{\displaystyle F \quad H}{\cdot|\cdot \quad \cdot|\cdot}}{F \div B \div N}} \div H$$

Here boron trifluoride is termed an acid because again a co-ordinate bond is formed.

While this concept can be extremely useful in considering, for example, certain organic reactions, it suffers from several disadvantages and need not concern us further here.

14.4. The strength of acids and bases

It has already been described (Section 14.1) how the strength of acids and bases may be given by the value of the dissociation constant K_a for the reaction

$$HA \rightleftharpoons H^+ + A^-$$

From what has been said in the previous Section it is obvious that the greater the ability of the solvent to act as a base, the greater will be the strength of the acid. In other words, the equilibrium in the reaction

$$HA + H_2O \rightleftharpoons H_3O^+ + A^-$$

will be well over to the right-hand side. The effective strength of any acid, therefore, can be varied simply by changing the solvent.

Acetic acid, for example, is normally thought of as a weak acid, i.e. a substance which has a slight tendency to donate protons. It can, however, be made to act as a proton acceptor, although it will not act in this way as readily as does water. The extent of the formation of the $CH_3CO_2H_2^+$ ion

$$\underset{\text{Acid}_1}{HA} + \underset{\text{Base}_2}{CH_3CO_2H} \rightleftharpoons \underset{\text{Acid}_2}{CH_3CO_2H_2^+} + \underset{\text{Base}_1}{A^-}$$

depends very much on the strength of the acid HA.

If we consider the strengths of some of the common acids in aqueous solution we find that a relationship exists between the strength and structure of the acids. The common acids are of two main types, *hydro acids*, (HCl, H_2S) and *oxy acids* (HNO_3, H_2SO_4).

In the hydro acid series, two generalizations can be made. Within any group in the periodic table, acid strength increases with increasing molecular weight. Thus acid strength increases in the group

$$HF < HCl < HBr < HI$$

Since it is the function of an acid to yield protons in solution this trend may seem somewhat anomalous when it is remembered that the $H-F$ bond has far more ionic character than has the $H-I$ bond. The important factor in this case is, however, the difference in size between the fluorine and iodine atoms. Because the iodine atom is so much larger than fluorine, the $H-I$ bond is correspondingly weak; consequently a proton can be more easily abstracted from HI by the action of a base.

The second generalization is that within any horizontal period of the periodic table, acid strength increases as the electro-negativity of the atom bonded to hydrogen increases. Thus in the series

$$NH_3 : H_2O : HF$$

HF is the strongest acid. This is because the change in size of the central atom in moving horizontally along a periodic series is not nearly so pronounced as that involved in moving down a group. Consequently the more polar the bond, the more readily may a proton be removed from the molecule.

Similarly the strength of an oxy acid is determined by both the electronegativity and the size of the central atom, the former usually being the dominant factor.

14.5. The pH scale

It has already been explained in Section 14.2 that water undergoes a self-ionization process to form hydrogen and hy-droxyl ions

$$H_2O + H_2O \rightleftharpoons H_3O^+ + OH^-$$

It is more convenient to write this equation as

$$H_2O \rightleftharpoons H^+ + OH^-$$

and simply assume that H^+ represents a hydrated proton. It is obvious from this equation that for each H^+ ion produced by the ionization of water, one OH^- ion is also produced. In other words, pure water contains an equal concentration of both H^+ and OH^- ions, and is said to be *neutral*.

The equilibrium constant for the ionization of water can be written

$$K = \frac{[H^+][OH^-]}{[H_2O]}$$

Since only a very small fraction of the water molecules ion-ize (at 25°C the concentration of both H^+ and OH^- ions is 1×10^{-7} g ions litre^{-1}) the concentration of undissociated water in the above equation may be taken as constant. By combining this constant concentration with the equilibrium constant a new constant K_w is obtained:

$$K[H_2O] = K_w = [H^+][OH^-]$$

where K_w is called the *ionic product* of water. In any aqueous solution, therefore, the product of the concentration of H^+ ions (g ions litre^{-1}) and the concentration of OH^- ions (g ions litre^{-1}) is a constant and has the value 1.0×10^{-14} at 25°C.

It follows from this that all aqueous solutions contain both H^+ and OH^- ions, even if the solution is strongly basic or acidic. For example, a 1 N solution of HCl, assuming complete dissociation, has a H^+ concentration of 1.0 g ion litre^{-1}. Since $K_w = [H^+][OH^-] = 1.0 \times 10^{-14}$ the concentration of OH^- ions is given by

$$[OH^-] = \frac{1.0 \times 10^{-14}}{[H^+]}$$

$$= 1.0 \times 10^{-14} \text{ g ions litre}^{-1}$$

Similarly, in a 1 N solution of NaOH, the OH^- concentration is 1.0 g ion litre^{-1}; thus the H^+ ion concentration is:

$$[H^+] = \frac{1.0 \times 10^{-14}}{[OH^-]}$$

$$= 1.0 \times 10^{-14} \text{ g ions litre}^{-1}$$

From these simple examples we find that in ordinary solutions the H^+ ion concentration can vary from 1.0 to 1.0×10^{-14} g ions litre^{-1}. Since this range of concentration is too large to be convenient, the acidity of a solution is usually expressed on a logarithmic scale called the *pH scale*. The pH of a solution is defined:

$$pH = -\log_{10}[H^+] = \log_{10}\frac{1}{[H^+]}$$

Thus in the examples already considered in this section, the pH may be calculated:

(a) Pure water

$$[H^+] = 1 \cdot 0 \times 10^{-7} \text{ g ions litre}^{-1}$$
$$pH = -\log_{10}(10^{-7})$$
$$= -(-7)$$
$$= 7$$

(b) 1 N HCl

$$[H^+] = 1 \cdot 0$$
$$pH = -\log_{10}(1 \cdot 0)$$
$$= 0$$

(c) 1 N NaOH

$$[H^+] = 1 \cdot 0 \times 10^{-14}$$
$$pH = -\log_{10}(10^{-14})$$
$$= -(-14)$$
$$= 14$$

It will be noticed that a pH value of 7 signifies a neutral solution, a value of less than 7 an acid solution, and a value of greater than 7 an alkaline solution.

14.6. Buffer solutions

There are occasions when it is essential to control the pH of a chemical system within a fairly narrow range, e.g. at certain stages in the manufacture of melamine-formaldehyde resins (see Chapter 27). This can be achieved by using a buffer solution, which is a solution whose pH is not appreciably altered by the addition of either H^+ or OH^- ions. Buffer solutions are usually mixtures of a weak acid and its salt.

The reason for the resistance to change in pH of a buffer solution can most readily be understood by considering a specific example. A mixture of acetic acid and sodium acetate can act as a buffer solution. If H^+ ions are added to such a system, the acetate ions will combine with the added H^+ ions to form undissociated acetic acid. Similarly, if OH^- ions are added, they will combine with the H^+ ions already in solution to form water, consequently some of the acetic acid molecules will then ionize to maintain the equilibrium. In neither case, therefore, will the pH of the system be altered to any extent.

Buffer solutions can be prepared to cover the whole pH range, but any given buffer solution is useful only over a range of about 2 pH units.

14.7. Common acids and bases

Sulphuric acid (H_2SO_4), perhaps the most well-known mineral acid, is widely used in the manufacture of fertilizers, textiles and paper. It is itself a fairly cheap chemical, the basic raw material being sulphur or a sulphur containing ore such as FeS_2 (pyrites). The sulphur is easily converted to gaseous SO_2 which may then be oxidized to sulphuric acid.

One method of achieving this last oxidation step is the *Lead Chamber Process*, where sulphur dioxide, oxygen and steam, together with oxides of nitrogen, are reacted in large lead-lined vessels. This is an example of a *homogeneous catalytic reaction*, where the nitrogen oxides catalyze the oxidation of the gaseous sulphur dioxide. The overall reaction may be represented:

$$SO_2 + H_2O \xrightarrow{NO_2} (NO)HSO_4 \xrightarrow{H_2O} H_2SO_4 + NO + NO_2$$
$$\text{nitrosyl sulphonic acid}$$

It will be noticed that the oxides of nitrogen react in the early stages of the reaction to form an intermediate compound which subsequently hydrolyses to form sulphuric acid, at the same time regenerating the nitrogen oxide catalyst.

The lead chamber process has now been superseded by the *Contact Process*, which is more suitable for the preparation of high purity and over-strength acids. In this case the oxidation of SO_2 to SO_3 is achieved by means of a *heterogeneous catalyst*. The original catalyst used was platinized asbestos, but vanadium pentoxide has been found to be cheaper and is not so susceptible to 'poisoning' by impurities. The overall reaction may be written:

$$SO_2 + O_2 \xrightarrow{V_2O_5} SO_3$$
$$SO_3 + H_2O \longrightarrow H_2SO_4$$

In practice it is found that the absorption of SO_3 in water is not very successful, but it is readily absorbed in sulphuric acid itself to yield pyrosulphuric acid, which can be converted to sulphuric acid by the addition of water.

$$SO_3 + H_2SO_4 \rightarrow H_2S_2O_7$$
$$H_2S_2O_7 + H_2O \rightarrow 2H_2SO_4$$

It is interesting to consider the contact process in terms of the principles of chemical equilibrium described in Chapter 13. We might expect that the reaction would be carried out at a high pressure since the formation of SO_3 from SO_2 and O_2 involves a reduction in volume. In fact the reaction is carried out at low pressure, the reason being that the rate of formation of SO_3 is determined by the rate of diffusion of the reactants to the catalyst surface through a strongly adsorbed layer of SO_3. Increasing the pressure does not accelerate this reaction. Furthermore, since the reaction

$$2SO_2 + O_2 \rightarrow 2SO_3$$

is exothermic, we expect that a low temperature will be necessary for maximum yield of product. Unfortunately the rate of reaction is found to be too slow at low temperatures and a compromise temperature of about 450°C is used.

Sulphuric acid is a strong dibasic acid, ionizing readily in aqueous solution in two stages:

$$H_2SO_4 \rightleftharpoons H^+ + HSO_4^-$$
$$HSO_4^- \rightleftharpoons H^+ + SO_4^{2-}$$

The first dissociation is complete, but the second occurs only to a limited extent. It is thus possible to make two types of salts from sulphuric acid: *acid salts* and *normal salts*.

$$H_2SO_4 + NaOH \rightarrow NaHSO_4 + H_2O$$
Sodium bisulphate

Sodium bisulphate is described as an acid salt because it contains the HSO_4^- ion which can itself dissociate to yield protons.

$$NaHSO_4 + NaOH \rightarrow Na_2SO_4 + H_2O$$

Apart from being used for the preparation of hydrochloric and nitric acids from their salts, sulphuric acid reacts with hydrocarbons to form another interesting series of acids called *sulphonic acids*.

These are monobasic acids, and their salts are important commercially as *detergents* (Chapter 28).

Concentrated sulphuric acid has a very strong affinity for water and indeed can form several hydrates, e.g. $H_2SO_4.H_2O$; $H_2SO_4.2H_2O$. An obvious application of this property is the use of H_2SO_4 as a drying agent or as a dehydrating agent, e.g. dehydration of alcohols.

Another important property of sulphuric acid is its ability to act as an oxidizing agent in concentrated solution. Many non-metals are readily oxidized by means of sulphuric acid, e.g. C, S, Br^-, I^-.

$$C + 2H_2SO_4 \longrightarrow 2SO_2 + CO_2 + 2H_2O$$

Furthermore, metals which lie below hydrogen in the electrochemical series, and which are therefore not attacked by dilute acids, can be readily oxidized by concentrated H_2SO_4

$$Cu + 2H_2SO_4 \longrightarrow CuSO_4 + 2H_2O + SO_2$$

Hydrochloric acid is the name commonly given to a solution of HCl in water, while pure HCl, which is a gas, is referred to simply as hydrogen chloride.

It has already been mentioned that HCl can be prepared by the action of sulphuric acid on metal chlorides, but it is also prepared as a by-product from the chlorination of organic materials, and this source is becoming increasingly important.

$$CH_3CH_3 + Cl_2 \longrightarrow CH_3CH_2Cl + HCl$$

Apart from undergoing the usual reactions with metals, with the evolution of hydrogen, hydrochloric acid in concentrated solution can act as a *complexing agent*. A very powerful solvent can be made by mixing three parts of concentrated hydrochloric acid with one part concentrated nitric acid. This mixture, called *aqua regia*, owes its high solvent power to the combination of the oxidizing properties of nitric acid with the complexing properties of the hydrochloric acid. Metallic gold, for example, which is insoluble in either of the acids alone, dissolves in aqua regia to form the complex anion $[AuCl_4]^-$.

Nitric acid, like HCl, can be simply prepared by the action of sulphuric acid on metal nitrates, but the bulk of commercial nitric acid is obtained from the hydrolysis of NO_2, which is in turn obtained by the oxidation of ammonia.

The high temperature oxidation of ammonia can yield two main products; nitrogen or nitric oxide (NO).

$$4NH_3 + 3O_2 \longrightarrow 2N_2 + 6H_2O$$
$$4NH_3 + 5O_2 \longrightarrow 4NO + 6H_2O$$

In the presence of platinum metal as catalyst, the second reaction is greatly accelerated relative to the first, and nitric oxide is the main product. This very readily reacts with oxygen to form nitrogen dioxide which dissolves in water to give nitric acid:

$$2NO + O_2 \longrightarrow 2NO_2$$
$$3NO_2 + H_2O \longrightarrow 2HNO_3 + NO$$

Nitric acid is a strong monobasic acid, being completely dissociated in dilute aqueous solution. It reacts readily with metals and metal oxides to form nitrates which are generally more readily soluble than the salts of other acids.

In the majority of its reactions, nitric acid acts as an oxidizing agent. In its reactions with metals, the gaseous product evolved is not usually hydrogen, but oxides of nitrogen. For example, a fairly non-reactive metal like copper will reduce nitric acid to NO_2 or NO, depending on the concentration of the acid used:

$$Cu + 4HNO_3 \xrightarrow[HNO_3]{conc.} Cu(NO_3)_2 + 2NO_2 + 2H_2O$$

$$3Cu + 8HNO_3 \xrightarrow[HNO_3]{dil.} 3Cu(NO_3)_2 + 2NO + 4H_2O$$

More reactive metals can reduce the nitric acid even further, to yield ammonium salts:

$$4Zn + 10HNO_3 \longrightarrow 4Zn(NO_3)_2 + NH_4NO_3 + 3H_2O$$

However, not all metals react readily with nitric acid. Some (e.g. Al, Cr, Fe) are rendered 'passive' by the formation of non-reactive oxide films on the metal surface. One further important application of nitric acid is its use in nitrating certain organic materials (Section 21.5).

Sodium carbonate (Na_2CO_3) is probably the most important alkali metal salt in industrial chemistry, being the cheapest available alkali. It is widely used in the manufacture of, for example, soap, glass and sodium hydroxide. The main method of manufacture is by the *Solvay process*, where CO_2 gas is bubbled through a solution of ammoniacal brine to form sodium bicarbonate which can readily be converted by heating to sodium carbonate. The reaction essentially involves the formation of HCO_3^- ions which will combine with the Na^+ ions in solution to form the sparingly soluble bicarbonate.

$$H_2CO_3 \rightleftharpoons H^+ + HCO_3^-$$

$$NH_3 + H^+ \rightleftharpoons NH_4^+$$

$$Na^+ + HCO_3^- \longrightarrow NaHCO_3$$

Carbonic acid is a weak acid which does not ionize to any appreciable extent. However, the presence of NH_3, which can combine with protons to form the NH_4^+ ion, greatly increases the ionization of the weak acid. The bicarbonate is converted to carbonate simply by heating, at the same time yielding CO_2 which can be recycled.

$$2NaHCO_3 \rightleftharpoons Na_2CO_3 + H_2O + CO_2$$

It should be noted that the other gaseous reactant (NH_3) can also be recycled, since it is easily recovered by treating the NH_4Cl by-product with lime

$$2NH_4Cl + Ca(OH)_2 \longrightarrow CaCl_2 + 2H_2O + 2NH_3$$

Sodium hydroxide is manufactured either by a *causticizing process* or by *electrolysis*. The former method involves heating a solution of Na_2CO_3 with a suspension of $Ca(OH)_2$, when Ca-CO_3 precipitates from solution

$$Na_2CO_3 + Ca(OH)_2 \rightleftharpoons 2NaOH + CaCO_3$$

Since both $Ca(OH)_2$ and $CaCO_3$ are solids, the equilibrium constant for the reaction is

$$K = \frac{[NaOH]^2}{[Na_2CO_3]}$$

from which it can be seen that the reaction is more complete in dilute solution. Since this would involve increased evaporation costs however, intermediate concentrations must be used.

The second method of manufacture involves the electrolysis of brine solution using a carbon anode and mercury cathode. Chlorine is evolved at the anode, and the sodium generated at the cathode dissolves in the mercury to form an amalgam which can then react with water in a separate cell to form sodium hydroxide solution.

$$2Cl^- \longrightarrow Cl_2 + 2e$$
$$Na^+ + 1e \longrightarrow Na$$

CHAPTER 15

Water Treatment

WATER is at the same time the most abundant and the most widely used chemical compound. Apart from being essential for life, it is necessary for a vast number of types of industry. It is, for example, the most common heat transfer agent and is used therefore as a coolant in condensers, cooling towers, etc. Other major uses are for steam generation and as a solvent. The overall importance of water may be appreciated when it is realized that the production of 1 ton of steel requires about 100 tons of water, while about 700 tons are required in the manufacture of 1 ton of paper.

15.1. Water sources

The primary source of all water is the sea, from which moisture is evaporated to be deposited later as rain. Sea water itself is sometimes used as a coolant, but because of the high salt content it generally cannot be used without extensive treatment. The large-scale purification of sea water is now becoming economically feasible and is discussed in some detail in Chapter 18.

For large-scale use, water may be obtained from one of two sources, *ground water*, i.e. water held underground in saturated zones of sand or porous rock, and *surface water*, i.e. rivers, lakes or artificial reservoirs. Water from these sources is never pure. Even rain water, the purest natural form of water, contains dissolved materials, e.g. dissolved oxygen, nitrogen and carbon dioxide, together with sulphur compounds if deposited over industrial regions. Once it reaches ground level even more impurities are gathered, depending on the nature of the terrain on which the water collects. Underground waters as obtained from wells are generally colourless and free of suspended material. They do, however, tend to have a fairly high concentration of dissolved solids, particularly calcium and magnesium bicarbonates and sulphates, and possibly iron salts. Surface waters can vary a great deal depending on locality. Moorland waters, for example, tend to be coloured brown and are often slightly acid due to the presence of weak organic acids derived from peaty ground. Lowland waters, on the other hand, tend to be colourless, but often contain suspended solids which settle only very slowly. Surface waters are almost inevitably contaminated by industrial or domestic effluents. It can readily be seen therefore that water may vary greatly according to its source, and will in general require treatment of some kind before it is suitable for industrial or domestic use.

15.2. Sterilization

Domestic water supplies may be put to various uses; the most important characteristics necessary are that it should be free of harmful bacteria and that it should be of pleasing appearance and taste.

Chemically the most common method of sterilization is *chlorination*. When chlorine is dissolved in water a chemical reaction takes place with the formation of hydrochloric acid, which is fully ionized, and hypochlorous acid, which dissociates to only a small extent.

$$Cl_2 + H_2O \rightleftharpoons H^+ + Cl^- + HOCl$$
$$\text{Hypochlorous acid}$$

$$HOCl \rightleftharpoons H^+ + OCl^-$$

These different species coexist in equilibrium, the relative amounts being determined mainly by the pH of the solution, and it is therefore immaterial whether the chlorine is introduced as such or as hypochlorite. In normal waters (pH 6–8) very little molecular chlorine exists and the bulk of the chlorine added is in the form either of undissociated hypochlorous acid or as the hypochlorite ion. It is generally believed that it is the undissociated acid which is the active species in the sterilizing reaction. The effectiveness of treatment depends both on the chlorine concentration used and on the time of contact, and different water supplies will of course require different doses. The ultimate test of effectiveness is a bacteriological examination of the water to ensure that sterilization is complete.

It is sometimes useful to add another type of chlorine-containing compound to the water supply, viz. *chloramines*, which may be formed by the action of either chlorine or hypochlorous acid on ammonia:

$$NH_3 + Cl_2 \rightleftharpoons NH_2Cl + H^+ + Cl^-$$

$$NH_3 + HOCl \rightleftharpoons NH_2Cl + H_2O$$

The compound NH_2Cl is called monochloramine, but compounds containing even more chlorine may be obtained:

$$NH_2Cl + HOCl \rightleftharpoons NHCl_2 + H_2O$$
Dichloramine

$$NHCl_2 + HOCl \rightleftharpoons NCl_3 + H_2O$$
Trichloramine

These reactions are all reversible and the chloramines decompose slowly in water to yield the active hypochlorous acid. They thus act as a reserve of chlorine which can kill organisms which develop after the initial treatment.

If it is necessary to over-chlorinate the supply initially, the excess chlorine must be removed because too high a concentration imparts both taste and colour to the water. This can be achieved fairly simply by passing the water through a bed of activated charcoal, or by treatment with suitable reducing agents, e.g. sulphur dioxide or sodium thiosulphate.

$$SO_2 + Cl_2 + 2H_2O \longrightarrow H_2SO_4 + 2HCl$$
$$2Na_2S_2O_3 + Cl_2 \longrightarrow Na_2S_4O_6 + 2NaCl$$

Other chemicals, e.g. permanganates, hydrogen peroxide, or ozone, can be used effectively to bring about sterilization, but are in general more expensive treatments and are not widely used.

15.3. Clarification

Natural waters usually contain an amount of suspended material, organic and inorganic, which must be removed. This can be achieved by a three-stage process of *settlement*, *coagulation* and *filtration*.

Settlement is brought about simply by prolonged storage of the water concerned. In general it is found that a storage time of around 24 hours removes about 90 per cent of the suspended material, but the remaining 10 per cent requires a different treatment. This is so because the particles suspended in the water are of different shapes and sizes. The rate of sedimentation of a spherical particle may be calculated from the equation:

$$v = \frac{2r^2(D-d)g}{9\eta}$$

where v = velocity of the particle,
r = radius of the particle,
D = density of the particle,
d = density of water,
η = viscosity of water,
g = gravitational constant.

The rate of sedimentation is seen to be dependent upon the radius of the particle. Particles of diameter greater than 2 or 3 microns (1 micron = 10^{-6} metre) will settle in less than 24 hours, but particles much smaller than this would take weeks or even years to settle out.

These extremely small particles less than 1 micron in diameter are described as *colloidal* and are so small that they cannot be filtered out. The most important property of material in this finely divided state is that it possesses a very high surface area and as a consequence is electrically charged due to the adsorption of ions on the surface. Since the small particles all carry charges of like sign, either positive or negative, they do not coalesce to form large aggregates because of the repulsive forces acting between them. In order to bring about coagulation, therefore, we must destroy these forces of repulsion. This can be achieved by adding highly charged ions to the water, these being adsorbed on the surface of the particles thus neutralizing the existing charge so that coagulation may occur. Compounds commonly used as coagulants include aluminium sulphate, ferric sulphate and activated silica. Recently, synthetic polyelectrolytes have been used as coagulants or coagulating aids.

After the coagulation and settlement treatment the water is finally filtered. This may be carried out by passing the water through a bed of sand or anthracite. In a new filter the sand initially exerts a simple sieving action, but normally the growth of a biological 'mat' of living organisms occurs in the surface layers of the filter bed and this helps to effect a further purification of the water.

15.4. Water hardness

We know that natural waters contain a large number of dissolved impurities, and the presence of certain specific impurities leads to the water being termed *hard*. Hardness was originally defined as the 'consumption' of soap by the water, due to the presence of calcium or magnesium ions which react with soap to form insoluble scums. Although other metal ions have this same property, e.g. iron, barium, manganese, these are rarely present in sufficient amount to cause any concern, and in practice the degree of hardness is determined by the amount of calcium or magnesium present.

Hardness may be of two types, *temporary* or *permanent*. Temporary hardness is caused by the presence of bicarbonates of magnesium or calcium and may be destroyed by boiling, when the corresponding carbonates precipitate from solution.

$$2HCO_3^- \xrightarrow{\text{heat}} CO_3^{2-} + H_2O + CO_2$$

$$Mg^{2+} + CO_3^{2-} \longrightarrow MgCO_3$$

Permanent hardness, which is not removed by boiling, is due to the presence of soluble salts of calcium or magnesium other than the bicarbonates, e.g. sulphates.

Measurement of hardness

Originally the degree of hardness of a specific water was determined by titration with a standard soap solution. Small amounts of the soap solution were added to a known volume of water, the mixture being shaken after each addition. The 'endpoint' of the titration was taken as the point where a stable lather was produced. The *total hardness* was estimated as above, the permanent hardness by repeating the experiment using boiled water, and the temporary hardness by difference.

A more recent method involves titrating the water sample with a reagent which forms complexes with calcium and magnesium ions. The usual reagent is ethylenediaminetetraacetic acid (EDTA) which is normally used as the disodium salt:

This large anion *chelates* (i.e. complexes) the calcium and magnesium ions and effectively removes them from solution:

When all the calcium and magnesium ions have reacted with the EDTA, a suitable indicator changes colour.

Since different water samples are hard for different reasons, e.g. one may contain calcium bicarbonate while another magnesium sulphate, the degree of hardness is by convention described in terms of a concentration of calcium carbonate. Thus from a knowledge of the appropriate equivalent weights we know:

$$50 \text{ g } CaCO_3 \equiv 12 \text{ g } Mg \equiv 42 \text{ g } MgCO_3 \equiv 81 \text{ g } Ca(HCO_3)_2$$

Unfortunately although convention decrees that the degree of hardness should be expressed in terms of calcium carbonate, the actual concentration units vary from country to country, and indeed within some countries. Probably the most useful method is to express the concentration as parts per million (ppm), since 1 ppm is equivalent to 1 mg per litre. Degrees of hardness are often found expressed as grains per gallon:

$$1 \text{ grain per Imperial gallon} = 14 \cdot 3 \text{ ppm}$$
$$1 \text{ grain per U.S. gallon} = 17 \cdot 1 \text{ ppm}$$

Waters with a hardness of less than 100 ppm are generally considered to be *soft*, while very hard waters will have values in excess of 300 ppm.

Removal of hardness by precipitation

The most obvious method of removing dissolved salts from water is by distillation, and in certain circumstances this procedure may be economic. A detailed account of the principles involved in removing fairly high concentrations of dissolved salts from water by distillation and also by the use of membranes is given in Chapter 18. The two most widely used methods for the removal of hardness on a large scale are the *lime-soda process* and *ion exchange processes*.

Before we consider how the unwanted ions may be precipitated from solution, we must learn something about the factors which can affect the solubility of sparingly soluble salts. If we consider the situation when a sparingly soluble salt such as magnesium hydroxide is placed in water, we find that a very small amount of the material dissolves to form magnesium and hydroxyl ions in solution. The solution process rapidly stops, however, because the solution soon becomes saturated. At this point a state of equilibrium is attained which may be represented by the equation

$$Mg(OH)_2(s) \rightleftharpoons Mg^{2+}(aq) + 2OH^-(aq)$$

We can write an equilibrium constant for this reaction, and since we are considering a heterogeneous reaction (see Chapter 13) this would have the form

$$K = [Mg^{2+}][OH^-]^2$$

This constant is called the *solubility product* and is normally written K_{sp}. The value of this constant determines the extent to which magnesium hydroxide can dissolve at any given temperature. If the ionic product of the magnesium and hydroxyl ions exceeds the value of the solubility product then magnesium hydroxide will precipitate from solution. Similarly if the ionic product is less than the solubility product, more magnesium hydroxide will dissolve until the equilibrium value is attained.

It is easy to see the effect of pH on the solubility of magnesium hydroxide. If the pH is increased, the concentration of hydroxyl ions in solution is increased. Since the ionic product $[Mg^{2+}][OH^-]^2$ must remain a constant, it follows that the magnesium ion concentration must decrease, i.e. magnesium hydroxide will be less soluble the higher the pH of the solution.

The solubility product of calcium carbonate is derived in the same way, i.e.

$$CaCO_3(s) \rightleftharpoons Ca^{2+}(aq) + CO_3^{2-}(aq)$$
$$K_{sp} = [Ca^{2+}][CO_3^{2-}]$$

The effect of pH on the solubility of calcium carbonate is not immediately obvious from the equation given above. It must be remembered, however, that in solution the carbonate ion can undergo further reactions and a number of equilibria are possible, e.g.

$$H_2CO_3 \rightleftharpoons H^+ + HCO_3^-$$
$$HCO_3^- \rightleftharpoons H^+ + CO_3^{2-}$$

We can see, therefore, that when carbonate ions are introduced into a solution, some will react with water to form the bicarbonate ion, thus reducing the actual concentration of carbonate ions. This effect will be larger in acid than in neutral solutions. The concentration of carbonate ions in the solubility product expression refers only to the material existing in that actual form at equilibrium, i.e. if $1 \cdot 0$ g ion per litre of carbonate ion is introduced and under the experimental conditions half of this changes to bicarbonate, then the concentration of carbonate is only $0 \cdot 5$ g ions per litre.

TABLE 15.1. SOLUBILITY
PRODUCT VALUES

Salt	Solubility product at 18°C
$Mg(OH)_2$	9×10^{-12}
$MgCO_3$	1×10^{-5}
$Ca(OH)_2$	1×10^{-6}
$CaCO_3$	1×10^{-8}

We can now consider the precipitation of calcium and magnesium from hard water. The lime-soda process involves the addition of hydrated lime and sodium carbonate to the water in order to precipitate the unwanted ions. Temporary hardness is removed by lime:

$$Ca(OH)_2 + Ca(HCO_3)_2 \longrightarrow 2CaCO_3 + 2H_2O$$
$$Ca(OH)_2 + Mg(HCO_3)_2 \longrightarrow Mg(OH)_2 + Ca(HCO_3)_2$$

We find that the calcium ions are simply removed as insoluble calcium carbonate, while the magnesium ions precipitate as the hydroxide, but at the same time an equivalent amount of calcium ions is introduced to the solution. This in turn must be removed by the addition of more lime as shown in the former equation. The net result is that the complete removal of magnesium bicarbonate requires twice as much lime as does the removal of calcium bicarbonate. The reason for the different behaviour of the calcium and magnesium salts can be seen in Table 15.1 where we find that for calcium the carbonate is the more insoluble salt, but for magnesium the hydroxide is the more insoluble and will therefore precipitate first from solution.

Permanent hardness due to calcium is removed by the addition of the sodium carbonate, while any permanent hardness due to magnesium is removed by the lime as shown above.

$$CaSO_4 + Na_2CO_3 \longrightarrow CaCO_3 + Na_2SO_4$$

It may seem paradoxical that hardness can be removed by adding to the solution a salt containing calcium, and it is true that care must be taken in treatment of this kind otherwise the water could finish with a higher degree of hardness than it had initially. Normally a slight excess of lime is added because the solubilities of both calcium carbonate and magnesium hydroxide decrease as the pH increases, but the hardness cannot be completely removed by this method.

Removal of hardness by ion exchange

The principle on which this method is based is that some materials, insoluble in water, are capable of absorbing certain ions from solution, at the same time liberating an equivalent number of different ions. If therefore we can find a material which will remove the ions that cause hardness, e.g. magnesium and calcium, while liberating harmless ions, e.g. sodium, then water softening is possible.

The earliest ion-exchange materials were naturally occurring inorganic compounds, *zeolites*, which are complex alumino-silicates. These are composed of large three-dimensional structures containing sodium ions rather loosely bound to the large interlocking lattice. When water containing calcium or magnesium ions is passed through a bed of this material, ion exchange occurs between the solution and the zeolite, as shown by the following equations (the symbol Z represents the complex zeolite):

$$\left.\begin{array}{c} Ca^{2+} \\ Mg^{2+} \end{array}\right\} + Na_2Z \rightleftharpoons \left.\begin{array}{c} CaZ \\ MgZ \end{array}\right\} + 2Na^+$$

The harmful calcium and magnesium ions are thus removed and an equivalent amount of sodium ions introduced in their place. The exchange reactions are reversible and the zeolites can be regenerated by passing an excess of brine solution through the zeolite bed.

The purification of water by ion exchange can be taken a step further by using synthetic *cation* and *anion exchange* resins which exchange hydrogen and hydroxyl ions respectively. A typical cation exchange resin would consist of a highly cross-linked polymer (e.g. cross-linked polystyrene or phenol-formaldehyde) containing acidic substituents such as the sulphonic acid grouping $Z-SO_3H$. When hard water is passed through a resin of this type exchange reactions analogous to the zeolite reactions occur:

$$\left.\begin{array}{c} Ca^{2+} \\ Mg^{2+} \end{array}\right\} + ZH_2 \rightleftharpoons \left.\begin{array}{c} CaZ \\ MgZ \end{array}\right\} + 2H^+$$

Again the exchange is reversible thus allowing the original resin to be regenerated by treatment with acid.

Anion exchange resins are cross-linked polymer structures containing basic substituents such as quaternary ammonium groups $Z-NR_3^+$. In this case exchange reactions occur with the liberation of hydroxyl ions:

$$2ZNR_3^+OH^- + SO_4^{2-} \rightleftharpoons (ZNR_3^+)_2SO_4^{2-} + 2OH^-$$

If water is treated with both a cation and an anion exchanger, not only is softening achieved but all mineral ions are removed from solution, leaving a water of exceptionally high purity.

15.5.   Boiler feed water

Water which is to be used in modern high-pressure boilers must be very carefully treated because of the extremes of temperature and pressure encountered in day to day working. The two main problems to be overcome are the prevention of harmful scale formation and the prevention of corrosion.

It might be thought that if the water were softened as described in the previous section no problem of scale formation would arise. It must be remembered, however, that although water may be demineralized by synthetic ion exchangers this process is expensive to apply on a large scale to hard waters, and the precipitation processes do leave a certain amount of *residual hardness*. Although only a small amount of material remains in solution, two processes occur in boiler operation to bring about scale formation, viz. the solubility of most scale-forming salts decreases with increasing temperature, and the rate of evaporation in a modern boiler can be very high. In either case the solubility product is eventually exceeded and scale deposition may occur.

One method of internal treatment is to add sodium carbonate to the boiler water. In this case, if the concentration of calcium ions builds up due to evaporation, calcium carbonate is precipitated as a sludge in the boiler. It appears at first sight that this is precisely what we wanted to avoid, but the carbonate is precipitated as a sludge which is easily removed from the boiler, and not as a highly cohesive film as would occur with, e.g., calcium sulphate under the same conditions.

At this point we encounter conflicting effects of scale prevention and protection against corrosion. Sodium carbonate cannot be added to high pressure boilers because the carbonate ion reacts under these conditions to yield hydroxyl ions:

$$CO_3^{2-} + H_2O \rightleftharpoons 2OH^- + CO_2$$

High local concentrations of hydroxyl ions can lead to *caustic embrittlement*, a type of corrosion to which steel is susceptible, where cracking occurs in areas of high stress, e.g. around rivet holes. It is therefore dangerous to use the sodium carbonate treatment in high pressure boilers.

The addition of phosphates in the correct concentrations can solve many of the problems posed by the previous method. Calcium phosphate is quite insoluble and is precipitated in the form of a sludge which is easily removed from the boiler. Unlike the carbonate ion, the phosphate is not attacked by water. A series of phosphate salts may form in solution depending on the pH, e.g.

$$H_2PO_4^- \rightleftharpoons HPO_4^{2-} \rightleftharpoons PO_4^{3-}$$
$$\xrightarrow{\text{increasing pH}}$$

Indeed, mixtures of phosphates act as buffer solutions (Section 14.6) thus helping to maintain the boiler water at a fixed pH.

Although simple phosphates may be added to the boiler water supply, it is sometimes more satisfactory to use polyphosphates with the general formula $(NaPO_3)_n$. These have a two-fold action in that they can be hydrolysed to form the simple phosphate ions discussed above, and have the property of forming soluble complexes with calcium ions, thus preventing the deposition of calcium salts in the boiler feed lines.

CHAPTER 16

Separation Techniques: I. Distillation

DISTILLATION is the process by which vapour is first formed from a liquid (through boiling), then condensed back to liquid (by cooling). The technique is used extensively in industry to achieve a separation of volatile from non-volatile material (as in the desalination of sea-water), and in the separation of a mixture of volatile components (as in the fractionation of crude petroleum).

16.1. Boiling-point diagrams

To acquire an understanding of the distillation process, let us turn our attention to the simplest type of binary (two-component) liquid mixture, namely, one in which the two components are miscible in all proportions. If the two components A and B are of a similar chemical constitution the system will deviate only slightly from ideality, and will have a boiling point intermediate between those of the pure components. The variation of boiling point with composition is shown by the lower curve (L) in Fig. 16.1, while the upper curve (V) gives the composition of vapour in equilibrium with liquid at the boiling point of the latter. The horizontal lines connecting the L-curve and V-curve give the compositions of liquid and vapour in equilibrium with each other at the appropriate temperature. Such a line is called a tie-line.

The reader should find it helpful in the understanding of the above diagram to consider the fate of a liquid mixture of composition l_1 as it is heated at constant pressure from temperature T_0 to temperature T_4. Until temperature T_1 is reached, the

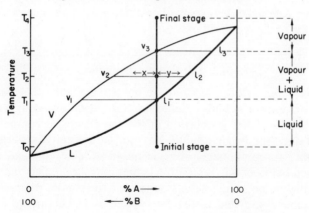

FIG. 16.1. Boiling point – composition diagram.

system is in the liquid state. At temperature T_1 the liquid boils producing vapour of composition v_1. This vapour contains a higher proportion of the more volatile component B than does the liquid with which the vapour is in equilibrium. This statement holds true for all liquid mixtures of the completely miscible type. As the temperature is further increased to T_2, a number of changes occur in the system. Firstly, the composition of the liquid changes along the L-curve to l_2; concurrent with this is the change in vapour composition along the V-curve to v_2. Secondly, the relative amounts of liquid and vapour change. Whereas at T_1, the amount of vapour v_1 is infinitesimally small, at T_2, the amount of vapour v_2 is comparable to the amount of liquid present. The relative amounts of the two phases can be determined from the distances x and y. Thus, for a system having an overall composition l_1,

$$\frac{\text{amount of liquid phase}}{\text{amount of vapour phase}} = \frac{x}{y}$$

As the temperature continues to increase, the above ratio becomes smaller and smaller, until at the temperature T_3 it reduces to zero, thus indicating that the system is now entirely in the vapour state. No further change of state occurs upon heating to the temperature T_4.

16.2. Fractional distillation

From our remarks so far, we have established that partial vaporization of a binary liquid mixture yields a vapour which contains a higher proportion of the more volatile component than does the original liquid mixture; as a result a separation of the two components is made possible. The technique employed to effect this separation is that of *fractional distillation*. By this method, the liquid is boiled and the vapour produced is condensed to yield a distillate which is richer in the more volatile component than the original mixture. When this operation is performed a sufficiently large number of times a separation is achieved. Fortunately the process can be carried out automatically through the use of a *fractionating column*, the most common example of which is the "bubble-cap" type (Fig. 16.2). The unit shown is designed to operate continuously with the liquid being fed into the column at a rate just sufficient to balance the material loss from the column. Once such a column has been operating for some time, a steady-state condition is

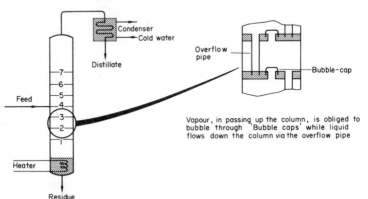

FIG. 16.2. Fractionating column.

attained, one characteristic of which is that a temperature gradient is established, the column being hot at the bottom, gradually cooling upwards.

Let us now consider the fate of a liquid feed of composition 50 per cent heptane in octane, as it is continuously introduced into the column at plate 4. Let the temperature of plates 1, 2, 3, ... be T_1, T_2, T_3, ..., etc. The feed introduced into plate 4 will assume a temperature T_4, at which temperature liquid of composition l_4 will be in equilibrium with vapour of composition v_4 (Fig. 16.3(a)). This vapour in rising up through the column is obliged to bubble through the liquid in plate 5 — in so doing its temperature drops to T_5 and liquid of composition l_5 condenses out. The vapour, now of composition v_5, rises, bubbles through liquid in plate 6 and partially condenses to liquid of composition l_6. The step-wise progression continues until at plate 7 the departing vapour is very nearly pure heptane. *Thus the vapour in rising up through the column experiences a continual enrichment in the more volatile component.*

Attention must now be drawn to the downward flow of liquid in the column (Fig. 16.3(b)). At plate 4, liquid of composition l_4 accumulates and overflows to plate 3 via the downspout. At the higher temperature T_3 partial vaporization of the more volatile component leaves a liquid of composition l_3 which in

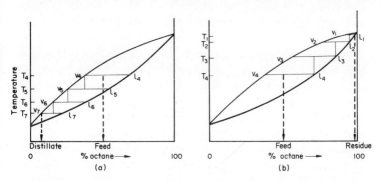

FIG. 16.3. Fractional distillation of a 50 per cent solution of heptane in octane.

turn overflows to plate 2 at which the composition of liquid is l_2. Finally a residue of almost pure octane is obtained, i.e. *the liquid in flowing down the column has become enriched in the less volatile component*. Thus, if a fractionating column has sufficient plates to allow efficient contact between ascending vapour and descending liquid, a complete separation of a liquid mixture of the type considered can be achieved.

16.3.   Petroleum refining

In the distillation considered above, the feed was a simple two-component mixture. When, however, a liquid mixture containing more than two components is being considered, a complete separation of each component becomes impossible if only one fractionating column is used. One particularly complex liquid mixture is crude petroleum, the first stage in the refining of which is to separate the mixture by distillation into fractions, each of which boil within a certain predetermined range. The petroleum is first heated to about 350°C, at which temperature it is almost completely vaporized. The vapour is then fed into the fractionating column (Fig. 16.4), where the pressure is maintained at a level slightly above atmospheric. By the mechanism already discussed, the more volatile materials condense in the upper plates of the column whilst the less volatile materials collect in the lower plates. The various fractions are continuously tapped off and in some cases processed by further distillation. Whereas the gasoline, kerosine and gas oil fractions can be distilled at near atmospheric pressure, the lubricating oil fraction is distilled under vacuum, thus reducing the boiling points of the oils in this fraction. This procedure is followed in order to avoid thermal decomposition of the oil, which becomes serious at temperatures in excess of about 400°C. To minimize still further the possibility of decomposition, vacuum distillation is frequently combined with steam distillation, which, for reasons to be discussed shortly, also has the effect of depressing the boiling point.

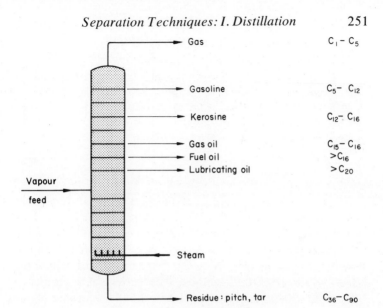

FIG. 16.4. Petroleum fractionation.

16.4. Constant boiling mixtures

Returning once more to the discussion of two component liquid mixtures, it must be mentioned that very marked deviations from ideal behaviour can arise. Such behaviour can be expected if the attraction between dissimilar molecules is either much greater or less than that between similar molecules; the former is illustrated by the system HNO_3–H_2O while the C_2H_5OH–H_2O system exemplifies the latter. Figures 16.5(a) and 16.5(b) show the type of boiling point diagrams for the above liquid pairs. An examination of Figs. 16.5(a) and 16.5(b) reveals that liquid mixtures corresponding in composition to points M and N, on boiling, yield vapours of exactly the same composition M and N. Such mixtures therefore boil unchanged and have constant boiling points — they are called *azeotropes* (which means to boil unchanged). Distillation of such mixtures

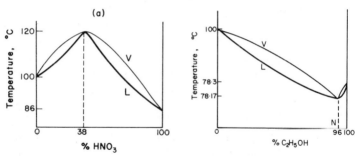

FIG. 16.5. Boiling point – composition diagrams for systems exhibiting a maximum and a minimum boiling point.

cannot result in a separation of the two components. Thus if a nitric acid solution of composition $100-38$ per cent HNO_3 is distilled, the distillate obtained is pure nitric acid, whilst the residue is the azeotropic mixture. Alternatively if the composition of the original solution was more dilute than that of the azeotropic mixture, the resulting distillate would be pure water. The products obtained upon distillation of HNO_3-H_2O and $C_2H_5OH-H_2O$ mixture are summarized in Table 16.1.

TABLE 16.1. DISTILLATION PRODUCTS OF HNO_3-H_2O AND $C_2H_5OH-H_2O$ MIXTURES

Original solution	$100-38\%$ HNO_3	$38-0\%$ HNO_3	$100-96\%$ C_2H_5OH	$96-0\%$ C_2H_5OH
Residue	Azeotrope	Azeotrope	C_2H_5OH	H_2O
Distillate	HNO_3	H_2O	Azeotrope	Azeotrope

16.5. Steam distillation – completely immiscible liquids

If two liquids are completely immiscible in each other there will be virtually no attractive interaction between the dissimilar molecules. As a result, the physical properties of each will remain unaltered even when mixed; they will exert their own vapour pressures independently. Thus, if $P_A°$ and $P_B°$ are the vapour pressures of the pure liquids A and B at a temperature T, then the partial pressures of A and B in the vapour phase above the liquid mixture will likewise be $P_A°$ and $P_B°$ respectively at temperature T. The total pressure P exerted by the vapour mixture will then be given by Dalton's law, viz.

$$P = P_A° + P_B°$$

This expression contains no terms which relate to the composition of the original liquid mixture, and so it follows that the total pressure is dependent only on the temperature and not at all on the relative amounts of the two liquids present. The total pressure at any temperature can be easily obtained by plotting the sum of the vapour pressures against the temperature (Fig. 16.6).

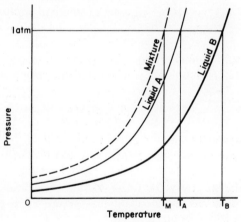

Fig. 16.6. Variation of vapour pressure with temperature for a liquid mixture made up of two completely immiscible liquids.

Now when any liquid system is heated under normal atmospheric pressure, it will boil when its vapour pressure equals that of the confining atmosphere, viz. 1 atm. An inspection of the graphs in Fig. 16.6 shows that this condition is satisfied at temperature T_M for the mixture, at temperature T_A for liquid A and at temperature T_B for liquid B. It is thus seen that *a liquid mixture comprising two immiscible liquids will boil at a temperature lower than the boiling point of either liquid and will continue to boil at that temperature (irrespective of its composition) provided the external pressure remains constant.*

The composition of the vapour produced from the boiling liquid also remains constant as the following procedure shows:

Let P_A° and P_B° be the vapour pressures of liquids A and B at the boiling point of the liquid mixture T_M. Then from Dalton's law

$$P_A^\circ = x_A P \quad \text{and} \quad P_B^\circ = x_B P$$

where x_A and x_B are the mole fractions of A and B in the vapour, and P is the total pressure = confining pressure.

$$\therefore \qquad \frac{P_A^\circ}{P_B^\circ} = \frac{x_A}{x_B} = \frac{n_A}{n_B} = \frac{w_A}{M_A} \bigg/ \frac{w_B}{M_B}$$

$$= \frac{w_A}{(w_A + w_B)M_A} \bigg/ \frac{w_B}{(w_A + w_B)M_B}$$

$$= \frac{WM_B}{(1-W)M_A}$$

where W = weight fraction of A in the distillate. Now M_A and M_B, being the molecular weights of A and B respectively, are constants; so also are P_A° and P_B° provided the temperature remains constant throughout the distillation as indeed it does. It therefore follows that the distillate composition expressed by W must likewise remain constant.

The term *steam distillation* is used when one of the two immiscible liquids is water, and the technique is employed in the distillation of the heavy oil fractions in petroleum refining in order to avoid the serious decomposition that would occur if the distillation took place at normal pressure. Steam distillation is thus seen to be, in a sense, equivalent to distillation *in vacuo*. The following example emphasizes this point.

EXAMPLE 16.1. *A heavy oil (mol. wt. = ca. 268) on steam distillation boils at a temperature t°C (where t < 100) under normal atmospheric pressure. The composition of the distillate is 99 per cent (by weight) of water. Calculate the pressure necessary to distil off the oil at the same temperature t°C using the technique of vacuum distillation.*

$$\frac{P^{\circ}_{H_2O}}{P^{\circ}_{HC}} = \frac{WM_{HC}}{(1-W)M_{H_2O}}$$

where $P^{\circ}_{H_2O}$ = vapour pressure of water at $t°C$,

P°_{HC} = vapour pressure of the hydrocarbon at $t°C$,

M_{H_2O} = molecular weight of water,

M_{HC} = molecular weight of the hydrocarbon,

W = weight fraction of water in the distillate.

Now P = total pressure

= atmospheric pressure

= $P^{\circ}_{H_2O} + P^{\circ}_{HC}$

$\therefore$
$$\frac{P - P^{\circ}_{HC}}{P^{\circ}_{HC}} = \frac{WM_{HC}}{(1-W)M_{H_2O}}$$

$\therefore$
$$P^{\circ}_{HC} = \frac{(1-W)M_{H_2O} \cdot P}{WH_{HC} + (1-W)M_{H_2O}}$$

$$= \frac{(1-0 \cdot 99) \times 18 \times 760}{0 \cdot 99 \times 268 + 0 \cdot 01 \times 18}$$

$$= 5 \text{ mm Hg}$$

Thus at $t°C$, the heavy oil exerts a vapour pressure of 5 mm Hg. Consequently, if it were to be distilled under vacuum at $t°C$, the external pressure would have to be reduced to 5 mm Hg.

CHAPTER 17

Separation Techniques: II. Extraction

DISTILLATION, considered in Chapter 16, can be thought of as an extraction technique. As such it relies on and exploits differences in the boiling points of liquids. However, in a number of situations distillation is a method ineffective for separation. This is so if the components making up the mixture have similar boiling points or if azeotropes are formed or if excessive decomposition occurs even at very low pressures. In such an event, solvent extraction (in which solubility differences are exploited) is normally attempted. This technique is one whereby a solute is transferred from one liquid in which it is initially dissolved, into another liquid immiscible in the first.

17.1. The distribution law

When a solute is added to two immiscible liquids, the solute distributes itself between the two solvents in such a manner as to make the ratio of its concentration in the two solvents a constant.† This is a statement of the *distribution law*, the ratio itself being called the *distribution coefficient (K)*. The latter has a value characteristic of the system in question and ideally is dependent only on the temperature.

Thus, if a solute S is added to two immiscible solvents A and B, then

$$K = \frac{[S]_A}{[S]_B}$$

†This is true provided the solute is in the same molecular state in each solvent.

where $[S]_A$ and $[S]_B$ represent the equilibrium concentrations of solute in solvents A and B respectively.

If an excess of solute is added both solvents must become saturated simultaneously in order that K should maintain its value appropriate to a particular temperature. Consequently the distribution coefficient is equivalent to the ratio of the saturated solubilities of solute in each of the two solvents.

17.2. Batch extraction

In extraction work, the two solvents are referred to as the *extraction solvent* and the *raffinate solvent*, the latter being the one in which solute is initially dissolved. At any stage in the extraction, the phase rich in the extraction solvent is termed the *extract phase*, while the raffinate-rich phase is labelled the *raffinate phase*.

In batch extraction, the extraction solvent is mixed with the raffinate phase until the equilibrium amount of solute is present in each case, after which the two phases are separated. The operation can then be repeated using a fresh batch of extraction solvent for each single extraction.

Suppose then that we have Wg of solute S dissolved in x litres of raffinate solvent. To this there is added y litres of extraction solvent. After mixing, the solute will be distributed between the two solvents and this distribution will be characterized by a particular value of the distribution coefficient which will then be given by:

$$K = \frac{[S]_{\text{Raffinate solvent}}}{[S]_{\text{Extraction solvent}}}$$

Let W_1 g = weight of solute remaining in the raffinate phase and $(W - W_1)$g = weight of solute transferred to the extract phase.

$$K = \frac{W_1/x}{(W - W_1)/y}$$

∴

$$W_1 = W \cdot \frac{Kx}{Kx + y}$$

The two layers are separated and to the raffinate phase there is added a fresh portion of y litres of extraction solvent. Again, on mixing, solute will pass from one phase into the other until the two concentrations are such that K assumes its appropriate value. Thus, if W_2 g is the weight of solute remaining in x litres of the raffinate phase after two extractions, $(W_1 - W_2)$ g is the weight of solute transferred to the second portion of extraction solvent. In this case

$$K = \frac{W_2/x}{(W_1 - W_2)/y}$$

$\therefore$
$$W_2 = W_1 \cdot \frac{Kx}{Kx + y}$$

$\therefore$
$$W_2 = W \cdot \left(\frac{Kx}{Kx + y}\right)^2$$

For n extractions, the weight W_n of solute remaining in the raffinate phase is given by:

$$W_n = W \cdot \left(\frac{Kx}{Kx + y}\right)^n$$

The process is described in the form of a flow diagram in Fig. 17.1.

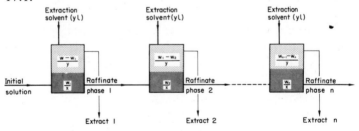

Fig. 17.1. Batch extraction.

EXAMPLE 17.1. *An aqueous solution of acetic acid was contaminated with small amounts of impurities and it was considered desirable to extract the acid with chloroform from which* *the acid could later be recovered by distillation. If the concentration of acid in the aqueous solution was initially 5 g l.$^{-1}$, calculate the minimum number of extractions required to reduce the concentration below 0·5 g l.$^{-1}$, using in each extraction 100 cm^3 of aqueous solution and 1 l. of chloroform.*

$$W_n = W\left(\frac{Kx}{Kx+y}\right)^n$$

$$\therefore \qquad n = \frac{\log Wn/W}{\log Kx/(Kx+y)}$$

$$K = \frac{[\text{Acetic acid}]_{H_2O}}{[\text{Acetic acid}]_{CHCl_3}} = 26\cdot4$$

Weight of acid in 100 cm^3 of aqueous solution $= 0\cdot5$ g $= W$.
Weight of acid in 100 cm^3 of aqueous solution $= 0\cdot05$ g after n extractions

$$\therefore \quad n = \frac{\log 0\cdot05/0\cdot5}{\log 26\cdot4 \times 0\cdot1/26\cdot4 \times 0\cdot2 + 1} = \frac{\log 0\cdot1}{\log 0\cdot7252} = 7\cdot163$$

$\therefore$Eight separate extractions are necessary to reduce the concentration of acid below 0·5 $g\,l.^{-1}$.

17.3.   Extraction efficiency

A measure of extraction efficiency is given by the ratio of weight of solute extracted to the initial weight of solute,

i.e.   per cent extraction $= \dfrac{100(W - W_n)}{W} = 100\left(\dfrac{1 - W_n}{W}\right)$

For a single extraction, where $n = 1$

$$\text{per cent extraction} = 100\left[1 - \left(\frac{Kx}{Kx+y}\right)\right] = \frac{100y}{Kx+y}$$

EXAMPLE 17.2. *The distribution coefficient of lactic acid between water and chloroform at 25°C is*

$$K = \frac{[\text{lactic acid}]_{H_2O}}{[\text{lactic acid}]_{CHCl_3}} = 49\cdot26$$

What percentage of lactic acid will be extracted from 50 cm^3 of an aqueous solution of the acid, by shaking with 1 l. of chloroform?

For $n = 1$, $x = 0.05$ l., $y = 1$ l.

$$W_1 = W\left(\frac{49.26 \times 0.05}{49.26 \times 0.05 + 1}\right)$$

$$= 0.7112W$$

$\therefore$ per cent extraction $= \dfrac{100(W - W_1)}{W}$

$$= \frac{100(W - 0.7112W)}{W} = 28.9 \text{ per cent}$$

or more directly:

per cent extraction $= \dfrac{100y}{Kx + y} = \dfrac{100 \times 1}{49.26 \times 0.05 + 1} = 28.9$ per cent

In the equation

$$W_n = W\left(\frac{Kx}{Kx + y}\right)^n$$

y and n are related to each other by the simple equation

$$y \times n = V$$

where $V =$ the total volume of extraction solvent.

We must now establish the relationship between extraction efficiency and the number of extractions. Thus for n extractions

$$\text{per cent extraction} = 100\left(1 - \frac{W_n}{W}\right)$$

$$= 100\left[1 - \left(\frac{Kx}{Kx + y}\right)^n\right]$$

$$= 100\left[1 - \left(\frac{Kx}{Kx + V/n}\right)^n\right]$$

$$= 100\left[1 - \left(\frac{1}{1 + V/Kxn}\right)^n\right]$$

The binomial expansion of $(1 + V/Kxn)^n$ produces the following series:

$$\left(1 + \frac{V}{Kxn}\right)^n = 1 + n\left(\frac{V}{Kxn}\right) + \frac{n(n-1)}{2}\left(\frac{V}{Kxn}\right)^2 + \cdots \left(\frac{V}{Kxn}\right)^n$$

When $n = 1$, $\left(1 + \dfrac{V}{Kxn}\right)^n = 1 + \dfrac{V}{Kx}$

Now this quantity is equivalent to the first two terms in the general series. From this it follows that

$$\text{for } n = 1, \quad 1 + \frac{V}{Kx} < \left(1 + \frac{V}{Kxn}\right)^n, \quad \text{for } n = 2,3,4; \cdots$$

Hence the extraction efficiency increases with the number of extractions. In other words, given a total volume V of extraction solvent, greater efficiency is obtained when n is made large and y small.

EXAMPLE 17.3. *Using the data given in Example 2 calculate the percentage of lactic acid which would have been extracted from 50 cm^3 of an aqueous solution of the acid, if two extractions had been performed with 500 cm^3 of chloroform per extraction.*

For $n = 2$, $x = 0.05$ l, $y = 0.5$ l.

$$W_2 = W\left(\frac{49\cdot26 \times 0\cdot05}{49\cdot26 \times 0\cdot05 + 0\cdot5}\right)^2$$

$$= 0\cdot6911\,W$$

$$\text{per cent extraction} = 100\left(\frac{W - W_2}{W}\right) = 30\cdot9 \text{ per cent}$$

Comparison of this figure with the corresponding figure obtained in Example 2 verifies the fact that extraction efficiency increases with the number of extractions.

17.4. Continuous countercurrent extraction

Batch extraction is a technique commonly employed in the laboratory but it finds little industrial application because of the inefficient use of the extraction solvent. The overall efficiency of the process can be greatly improved by the technique of *continuous countercurrent extraction*. By this method, the extraction solvent is made to flow continuously in one direction, the raffinate phase flowing continuously in the reverse direction. Now whereas in batch extraction mixing and settling take place in the same vessel, two separate vessels (referred to as the mixer and settler) are required when the flow of the liquids is continuous. Thus, in the two-stage countercurrent extraction shown in Fig. 17.2, fresh extraction solvent is used in the second

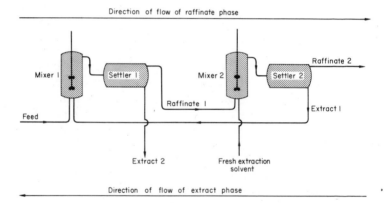

Fig. 17.2. Continuous countercurrent extraction.

extraction, the extract then being used in the first stage. For clarity only two stages have been considered; usually more stages are necessary. Although the extraction efficiency of this technique is high, the floor area required for such a multistage extraction is likewise high. This disadvantage is eliminated by using vertical tower extractors, one of the most modern of which is shown in Fig. 17.3.

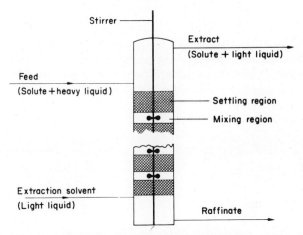

Fig. 17.3. Vertical tower extractor.

The process is basically a continuous countercurrent extraction with the more dense liquid being fed in at the top of the column, the less dense liquid entering at the bottom. There is a continuous transfer of solute material (which can be initially contained in either the light or heavy liquid) as one rises up and the other flows down the column. The column itself is essentially an assembly of mixers and settlers, the mixing being provided for by a rotating stirrer, while settling is achieved through the use of wire mesh.

17.5. Industrial applications

The technique of solvent extraction is most extensively used in the petroleum industry. For example, in the refining of the lubricating oil fractions the constituents of the oil are distributed between two immiscible solvents, e.g. propane and a phenolic mixture of phenols, cresols and xylenols. The type of plant used is similar to that shown in Fig. 17.2 in that it is made up of a series of mixer-settlers. However, in this case the oil is fed not into the first mixer, but into an intermediate one. The process is shown in its most simplified form in Fig. 17.4. Propane is introduced at one end of the series of mixer-settlers, and as it moves along the series, preferentially dissolves out the alkanes. In the meantime the phenolic solvent, introduced at the other end of the series and flowing countercurrently to the propane, dissolves out the cycloalkanes and aromatics.

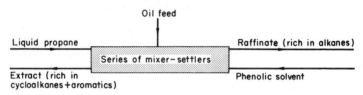

FIG. 17.4. Refining of the lubricating-oil fractions.

Another application is the removal of aromatics from kerosine using liquid sulphur dioxide. In this case the liquid sulphur dioxide is fed into the top of a column with the less dense kerosine entering at the bottom. The kerosine in floating to the top of the column makes intimate contact with a countercurrent flow of liquid sulphur dioxide and, in so doing transfers its aromatic constituents to sulphur dioxide.

Although being particularly effective in removing aromatics, sulphur dioxide suffers from the disadvantage that it is a gas at normal pressures. Consequently the extraction unit must be designed to operate under pressure. To overcome this other less volatile solvents are sometimes used.

In addition to extracting aromatic compounds, sulphur dioxide is also effective in reducing the sulphur content of oil fractions. It is, however, very much less efficient than liquid hydrogen fluoride which is now commonly used for this purpose.

CHAPTER 18

Separation Techniques: III. Desalination

THE processes of evaporation and condensation occur on a vast scale over the earth's surface. Where the surface is covered by water, evaporation occurs and the water vapour produced is carried by winds over land where it condenses as rain or snow.

The advent of the industrial revolution brought about a concentration of population in the rapidly growing cities and necessitated the construction of large reservoirs in order to trap condensed water. Today, however, the demand for fresh water is increasing at such a rate that these measures are becoming more and more inadequate. Even in the U.K., with its temperate climate, water shortages are not uncommon. The situation in many countries is more critical than this; deserts must be reclaimed, but in a country where the rainfall is slight and the nearest fresh water supply is perhaps a thousand miles distant, this is no mean task. A partial solution is provided by the improvement and extension of existing trapping facilities, but in recent years much research and development has been concentrated on the sea as a source of fresh water. Unfortunately, the sea contains about 3·5 per cent of salts (mostly sodium chloride with much lesser amounts of magnesium chloride, and magnesium and calcium sulphates). Small though this figure may seem, it is intolerable from the standpoint of human or agricultural needs – it is indeed too great by a factor of about 70.

A large number of processes, designed to extract fresh water from the sea and referred to collectively as *desalination processes*, have been suggested. Of these, distillation and freezing processes and reverse osmosis are the furthest advanced and our discussion will be restricted to these techniques.

18.1. Multistage flash distillation

The separation of fresh water from the sea is a simple enough matter, easily accomplished by, for example, distillation. The problem, however, is not *how* the separation can be achieved but *how efficiently* it can be done. A simple distillation, such as is shown in Fig. 18.1, would produce fresh water as distillate but would do so most inefficiently in that (i) the process is batch as distinct from continuous, and (ii) there is no recovery of the latent heat released when the vapour condenses in the condenser. If the technique of distillation is to be employed to obtain fresh water from the sea, then in order to be economically viable, there must be a recovery of this latent heat and at the same time the plant must be designed to allow a continuous flow of sea water. These requirements are particularly well satisfied in the technique referred to as *multistage flash distillation*.

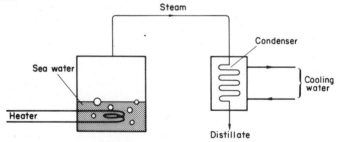

Fig. 18.1. Simple distillation.

By this method, sea water at a certain temperature is introduced into a vessel (the flash chamber) in which the pressure is maintained at a value less than that corresponding to the vapour pressure of the sea water. This condition results in rapid evaporation, the vapour being said to 'flash' off. Now before considering the multistage process, let us examine the essential features of a single stage flash distillation, as shown in Fig. 18.2. A sea-water feed enters the condenser tubes and, immediately prior to its entry into the flash chamber, is heated by steam to a temperature T at which its vapour pressure is P. It then enters the flash chamber where the pressure is maintained at P_1 (where $P_1 < P$). This condition causes a fraction of the liquid to flash off and because it is only the more energetic water molecules which leave the solution, the latter will experience a cooling effect, the temperature dropping to T_1. In the meantime, the flashed vapour condenses on the condenser coils, thus releasing its latent heat of condensation, which, by heat transfer through the walls of the condenser, is absorbed by the sea water.

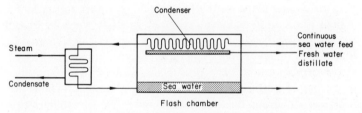

FIG. 18.2. Single-stage flash distillation.

A number of stages constitute the multistage process, which is outlined in a simplified form in Fig. 18.3. As can be seen, the sea-water feed entering the condenser in the nth flash chamber is preheated by condensing vapour; this recovery of energy is repeated in each subsequent condenser. The

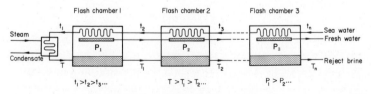

FIG. 18.3. Multistage flash distillation.

temperature is then raised, as before, by steam and this is followed by the flashing off of vapour in successive chambers. The salt water as it moves along from one chamber to the next cools continually and also experiences an increase in salt concentration. Because of this cooling effect the pressure in each successive chamber must progressively decrease in order that the pressure within the chamber is less than the vapour pressure of the feed to that chamber. The vapour, condensing out as a fresh-water distillate, could be withdrawn from each chamber. It is, however, more efficient to pump the distillate from one chamber to the next, for it also is cooled by flashing and, in this way, the preheating of the brine is made more appreciable.

The overall efficiency of a multistage flash distillation unit is largely dependent on the flashing temperature range $T-T_n$. Whilst the lowest temperature T_n is virtually fixed by that of the sea-water feed, the higher temperature T is determined by that of the steam supply. Thus if high-pressure steam is used, sea water could enter the first flash chamber at temperatures in excess of 100°C, thereby increasing efficiency. There are, however, inherent disadvantages in the use of temperatures much higher than this, the most serious of these being an increase in corrosion and the formation of scale (see Chapter 15). Consequently, present practice is to keep the high temperature T below 110°C. However, looking to the future, it seems likely that higher temperatures will be employed as corrosion and scale difficulties are overcome through the use of resistant materials and improved techniques.

18.2. Freezing processes

Distillation is not the sole technique which can bring about the separation of fresh water from the sea. The freezing of sea water separates out ice leaving the salts in solution. Once more, however, the efficiency of the process is of paramount concern. Two freezing processes, each based on a common principle, will be discussed. These are the *vacuum freeze process* and the *secondary refrigerant process*.

Vacuum freeze process

By this method pre-cooled sea water at about 1°C is fed into a chamber (the freezer) where the pressure is maintained at a value less than that corresponding to the vapour pressure of freezing sea water (about 3 mm Hg) in order that some water may flash off and produce a consequent cooling effect, sufficient to freeze out some ice (Fig. 18.4). Because of the low temperature and consequent low value of vapour pressure, the freezer chamber must operate under high vacuum. The ice–brine mixture is pumped to a separator, where it is both separated and washed with fresh water, the washing being necessary to remove brine trapped between the ice crystals. Concurrently the water vapour, initially produced by flashing from the freezer, is compressed and condensed on washed ice contained in the melter. Thus at this stage water is produced by condensation of vapour together with the melting of ice.

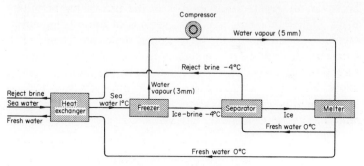

FIG. 18.4. Vacuum freeze process.

Secondary refrigerant process

Whereas the freezing of sea water is achieved in the vacuum freeze process by evaporating a portion of the water under high vacuum, this process accomplishes the freezing by vaporizing another liquid immiscible in water. The vapour of this second liquid is then compressed and used to melt the ice. This second liquid is referred to as the secondary refrigerant. The advantage of using this technique is that a secondary refrigerant can be chosen which boils under normal atmospheric pressure in the region of 0°C. *n*-Butane (b.p. 0°C) is ideal as a secondary refrigerant.

The process operates as follows. Liquid butane, introduced into the freezer, evaporates to a considerable extent; the energy required for this evaporation is extracted from the sea water which cools and produces, as before, an ice–brine mixture which is transferred to the separator and then on to the melter (Fig. 18.5). Simultaneously the butane vapour, removed from the freezer, is compressed and condensed on the ice in the melter thus regenerating liquid butane (which is then returned to the freezer) and producing fresh water (a small fraction of which is returned to the separator to wash the ice crystals).

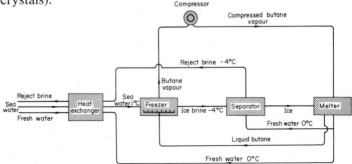

FIG. 18.5. Secondary refrigerant process.

An advantage of freezing processes over distillation techniques is their low energy requirement. However, practical difficulties associated with the separation and washing stages have up till now prevented the commercial exploitation of the processes although fairly large pilot plants are in operation.

18.3. Reverse osmosis

The phenomenon we now call *osmosis* was first noted by the Abbe Nollet in 1748. He observed that when wine was stored in a vessel sealed off with an animal membrane, such as a bladder, and immersed in water, the bladder swelled up and finally burst. This, we now know, is a result of the bladder permitting water (solvent) but not alcohol (solute) to pass through it. A similar effect can be seen when prunes or raisins swell and subsequently burst when immersed in water. Osmosis can thus be defined as *the spontaneous flow of solvent from a dilute to a more concentrated solution via a membrane which allows solvent but not solute molecules to pass through it.* Such a membrane is said to be *semipermeable*.

In Fig. 18.6, a tube fitted with a semipermeable membrane and containing a salt solution is partly immersed in water so that initially the two levels outside and inside the tube are the same. Water will tend to flow from the beaker into the tube through the semipermeable membrane. This flow can be prevented by applying a pressure of such a magnitude as to maintain the equality of the two levels. This excess pressure required to prevent osmosis is called the osmotic pressure, and is denoted by the symbol π. Now we can vary the pressure acting on the solution by incorporating a piston into the tube. Thus, if the flow of water is to be prevented, the excess pressure P acting on the piston must be equal to the osmotic pressure of the solution, π (Fig. 18.6(a)). If P is less than π, the piston will be pushed upwards as a result of the spontaneous flow of water into the tube via the semipermeable membrane. If no excess pressure is applied to the solution then water will enter the tube until the hydrostatic pressure exerted by the column of solution becomes equal to the osmotic pressure (Fig. 18.6(b)). Finally, if a pressure in excess of π is applied to the solution, water will pass *out of* the solution (Fig. 18.6(c)). This reversal of the normal spontaneous flow of solvent is referred to as *reverse osmosis*.

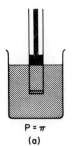 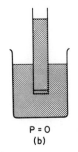 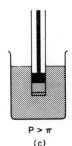

P = π   P = 0   P > π
(a)   (b)   (c)

FIG. 18.6. Osmotic flow of solvent.

As a technique for desalination, reverse osmosis has been under development for only a short time, and although the process is basically attractive (on account of its simplicity and low energy requirements), its commercial success will depend upon the production of membranes which are both cheap and durable.

Sea water has an osmotic pressure of about 25 atm and, if reverse osmosis is to occur, a pressure in excess of this must be applied on the solution side of the membrane (cellulose acetate). In practice it is found that much higher pressures than this must be developed and this necessitates the use of a membrane support. Fibre glass, on account of its low cost, is particularly suitable for this purpose. At present, development of the technique has not passed beyond the pilot plant stage and is, as yet, only successful in the recovery of fresh water from a brackish source, i.e. one in which the salt content ranges from a tenth to 1 per cent by weight.

CHAPTER 19

Organic Chemistry — Basic Concepts

19.1.   What is organic chemistry?

Organic chemistry is defined as *the chemistry of carbon compounds*. Other elements occur frequently — notably hydrogen, which is seldom absent, and to a rather lesser extent oxygen, nitrogen, sulphur and the halogens. The appearance in organic compounds of elements outside of this group is comparatively rare.

It is striking that one major branch of chemistry is devoted to the compounds of only one of the 103 elements now placed in the periodic table. The reasons for this we shall presently examine.

19.2.   Chemical bonding in carbon compounds

Carbon exhibits a valency of four. Moreover, the valence links formed by carbon are essentially covalent. The covalent nature of organic compounds results in minimal molecular interaction due to electrical charge, and most of these materials are gases, liquids or low melting solids. Organic compounds are usually non-electrolytes and generally insoluble in water or other ionizing solvents. Valence type controls physical properties in these and other ways.

A given carbon atom may be single bonded to other monovalent elements, as in methane or methyl chloride; or it may form multiple bonds with multivalent elements, as it does in formaldehyde and hydrogen cyanide.

$$
\begin{array}{cccc}
\text{H} & \text{H} & \text{H} & \\
| & | & \diagdown & \\
\text{H—C—H} & \text{H—C—Cl} & \quad\text{C}=\text{O} & \text{H—C}\equiv\text{N} \\
| & | & \diagup & \\
\text{H} & \text{H} & \text{H} & \\
\text{Methane} & \text{Methyl chloride} & \text{Formaldehyde} & \text{Hydrogen cyanide}
\end{array}
$$

In being tetravalent carbon is no different from other elements in group IV of the periodic table. It is capable of forming a uniquely large number of compounds (and so carbon chemistry is a uniquely large subject) because carbon has a great tendency to combine, not only with other elements, but also with itself.

Thus carbon in combination with hydrogen is potentially capable of forming, in addition to methane, a whole series of related *hydrocarbon* compounds such as ethane, propane, butane, and so on.

$$
\begin{array}{ccc}
\text{H} \ \ \text{H} & \text{H} \ \ \text{H} \ \ \text{H} & \text{H} \ \ \text{H} \ \ \text{H} \ \ \text{H} \\
| \ \ | & | \ \ | \ \ | & | \ \ | \ \ | \ \ | \\
\text{H—C—C—H} & \text{H—C—C—C—H} & \text{H—C—C—C—C—H} \\
| \ \ | & | \ \ | \ \ | & | \ \ | \ \ | \ \ | \\
\text{H} \ \ \text{H} & \text{H} \ \ \text{H} \ \ \text{H} & \text{H} \ \ \text{H} \ \ \text{H} \ \ \text{H} \\
\text{Ethane} & \text{Propane} & \text{Butane}
\end{array}
$$

The extension of carbon chains by the formation of *saturated* carbon–carbon single bonds is a well-established phenomenon of organic chemistry. Carbon atom may also be bonded to carbon atom by *double* or *triple* valence links. The parent hydrocarbon compounds with these features are the gases ethylene and acetylene. Structures containing double or triple bonds are termed *unsaturated*.

$$
\begin{array}{cc}
\text{H} \ \ \text{H} & \\
| \ \ | & \\
\text{C}=\text{C} & \quad \text{H—C}\equiv\text{C—H} \\
| \ \ | & \\
\text{H} \ \ \text{H} & \\
\text{Ethylene} & \text{Acetylene}
\end{array}
$$

In organic compounds typified by methane, ethylene and acetylene the carbon skeleton is in the form of an open chain.

Open chain compounds as a class are known as *aliphatic* compounds. Carbon skeletons can occur not only as open chains but also as rings. Thus *alicyclic* compounds (known also as *naphthenes*) such as the hydrocarbons cyclopropane and cyclohexane provide further examples of the versatility with which carbon atoms bond to one another. *Aromatic* compounds (discussed in Chapter 21) constitute a second very important class of ring structures in organic chemistry. The hydrocarbon benzene, C_6H_6, is the parent member of this class.

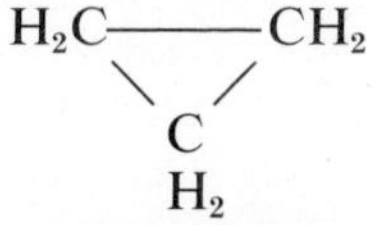

Cyclopropane

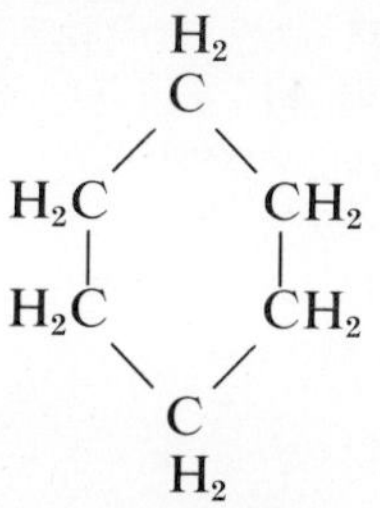

Cyclohexane

19.3. Structural isomerism

There is for all but the simplest organic structures more than one way of arranging the atoms involved while still satisfying valence requirements. Thus in the hydrocarbon series methane, ethane, propane, butane and beyond the first three are described by one unambiguous structural representation. For the fourth member, butane C_4H_{10}, however, two satisfactory structures, shown below, can be written out. These are most commonly called *normal-* or *n*-butane (the prefix *normal* being used to indicate a straight chain arrangement of carbon atoms) and isobutane.

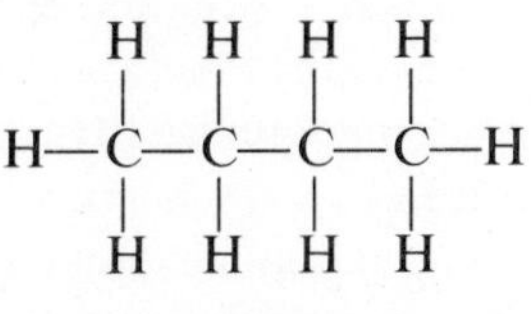

n-Butane

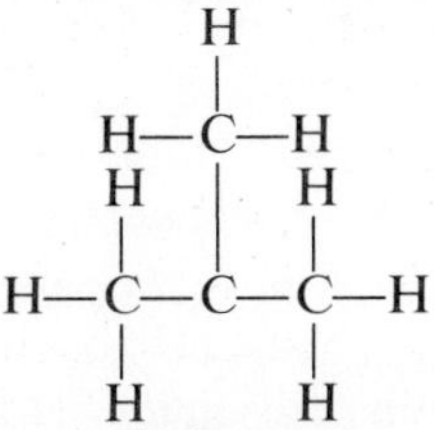

Isobutane

The property whereby a single molecular formula may represent more than one structure is known as *structural isomerism*. Butane can exist as one of two structural isomers. The next member of this particular hydrocarbon series, pentane C_5H_{12}, can exist as one of three structural isomers. Beyond this

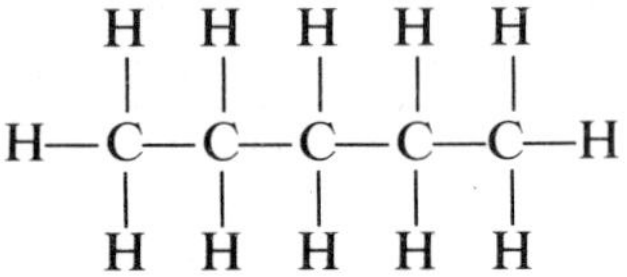

n-Pentane

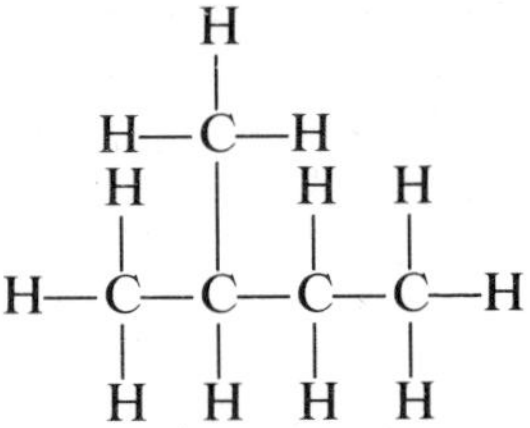

Isopentane

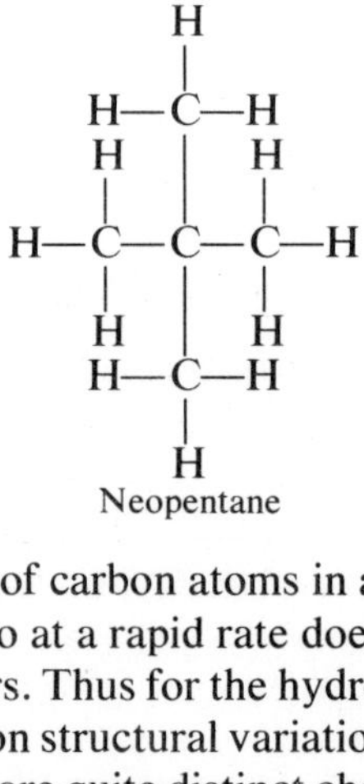

Neopentane

point, as the number of carbon atoms in a given molecular formula increases, so too at a rapid rate does the number of possible structural isomers. Thus for the hydrocarbon pentacosane, $C_{25}H_{52}$, over 36 million structural variations are possible.

Structural isomers are quite distinct chemical compounds. In a group of isomers like pentanes, however, there are only minor differences in chemical behaviour between isomers, as we might expect from the similar carbon–carbon and carbon–hydrogen valence bonds in all three compounds. Structural isomerism, however, often tends to lead to very major differences in chemical behaviour. Thus isomeric arrangements of the molecular formula C_3H_6O, shown below, contain in part quite different types of valence bonds from one another, and the two compounds in question, propionaldehyde and methyl vinyl ether, are easily distinguished chemically from one another.

Propionaldehyde   Methyl vinyl ether

19.4.   The homologous series

Table 19.1 lists the first eight members of the hydrocarbon family of compounds called alkanes. Only the molecular formulae are given, but remember that C_4H_{10} and higher formulae each represent more than one isomer.

TABLE 19.1. ALKANES

Name	Molecular formula
Methane	CH_4
Ethane	C_2H_6
Propane	C_3H_8
Butane	C_4H_{10}
Pentane	C_5H_{12}
Hexane	C_6H_{14}
Heptane	C_7H_{16}
Octane	C_8H_{18}

A series of this sort is called an *homologous series*. Inspection shows that each member differs from its predecessor by a constant amount (i.e. CH_2, or a molecular weight difference of 14, in our example). The above series can be expressed by the general formula C_nH_{2n+2}. Each individual member of the series is called an *homologue*.

Any homologous series shows a steady change in physical properties such as boiling point or specific gravity as the series is ascended. More important, all members of a given series exhibit very similar chemical behaviour. This permits the apparent complexity of organic chemistry, therefore, to be reduced to a study of a relatively small number of homologous series.

19.5. Functional groups

In most organic compounds there is present an atom or group of atoms whose characteristic chemical reactivity can be used as a basis for classification. This particular active point of a molecule is called the *functional group*. In a given homologous series of compounds, each member has present the same functional group.

In the alkane *n*-butane both carbon–hydrogen and carbon–carbon bonds may be broken in the course of chemical reaction and each of these sites is a potential functional group. In the hydrocarbon 1-butene, however, the carbon–carbon double bond is much more active than any other bond in the molecule, and is by definition the functional group. Again, if one of the hydrogen atoms in *n*-butane is replaced by a bromine atom, as in 2-bromobutane, subsequent chemical reactions are found to involve in the main the carbon–bromine bond: this site in the molecule is the chemically dominant one, and is therefore the functional group.

$$CH_3 \cdot CH_2 \cdot CH_2 \cdot CH_3 \qquad CH_2{=}CH \cdot CH_2 \cdot CH_3$$
$$\quad \text{\textit{n}-Butane} \qquad\qquad\qquad \text{1-Butene}$$

$$\overset{\displaystyle Br}{\overset{\displaystyle |}{CH_3 \cdot CH \cdot CH_2 \cdot CH_3}}$$
$$\text{2-Bromobutane}$$

19.6. Hydrocarbon radicals

Every organic structure may be regarded as having been derived from a hydrocarbon by substitution of the relevant functional group in place of one or more hydrogen atoms.

The fundamental position of a hydrocarbon homologous series can be demonstrated by reference to the alkane series (Table 19.1). If in the alkanes one hydrogen atom is replaced by one chlorine atom, the pattern shown in Table 19.2 results.

The monochlorinated alkanes obviously constitute an homologous series, the *alkyl chlorides*, fitting the general formula $C_nH_{2n+1}Cl$.

TABLE 19.2. MONOCHLORINATED
ALKANES

Name	Molecular formula
Methyl chloride	CH_3Cl
Ethyl chloride	C_2H_5Cl
Propyl chloride	C_3H_7Cl
Butyl chloride	C_4H_9Cl
	$C_nH_{2n+1}Cl$

If instead of chlorine we were to substitute some other monovalent atom or group X in the alkanes a series of general formula $C_nH_{2n+1}\cdot X$ would be obtained. The substituent group, which is in practice the functional group, may be complex, and may itself contain a carbon atom; e.g. substitution of the group

$$-C\overset{\displaystyle O}{\underset{\displaystyle H}{\big\langle}}$$

would yield the series of aliphatic *aldehydes* whose general formula is therefore $C_nH_{2n+1}\cdot CHO$.

The hydrocarbon residues present in these substituted alkanes are called *alkyl radicals*—the methyl radical from methane, the ethyl radical from ethane, and so on. They are denoted by the general symbol R; hence the general formula commonly used for an alkyl chloride is $R\cdot Cl$.

Alkanes, then, form the basis of a wide selection of different organic aliphatic compounds classified, according to functional group, in homologous series. A parallel relationship exists in aromatic chemistry. The precursor of all aromatic compounds, benzene, C_6H_6, may have one or more of its hydrogen atoms replaced by other atoms or groups. Thus the first member of an aromatic homologous series can be represented as $C_6H_5 \cdot X$; the hydrocarbon radical C_6H_5 is called the phenyl radical. Radicals derived from aromatic systems in general are called *aryl radicals*, and are denoted by the symbol Ar.

19.7. Radical isomerism

The monochlorination of either methane or ethane yields in each case one product only; propane, however, can be monochlorinated to give one of two structural isomers, *n*-propyl chloride or isopropyl chloride.

$$CH_3 \cdot CH_2 \cdot CH_2Cl \qquad \overset{\displaystyle Cl}{\underset{}{\overset{|}{CH_3 \cdot CH \cdot CH_3}}}$$

n-Propyl chloride   Isopropyl chloride

The propyl radical can therefore exist in one of two isomeric forms. In the case of butane there are two hydrocarbon isomers (*n*-butane and isobutane) from each of which can be derived two radical isomers. The four possible butyl chlorides are shown below.

$$CH_3 \cdot CH_2 \cdot CH_2 \cdot CH_2Cl \qquad \overset{\displaystyle Cl}{\underset{}{\overset{|}{CH_3 \cdot CH \cdot CH_2 \cdot CH_3}}}$$

n-Butyl chloride   *Sec*-Butyl chloride

$$\begin{matrix} CH_3 \\ \diagdown \\ \qquad CH \cdot CH_2Cl \\ \diagup \\ CH_3 \end{matrix} \qquad \overset{\displaystyle CH_3}{\underset{\displaystyle CH_3}{\overset{|}{CH_3 \cdot CCl}}}$$

Isobutyl chloride   *Tert*-Butyl chloride

The prefix *secondary* or *sec* preceding a radical name indicates attachment of a substituent group to a carbon atom to which is also attached one hydrogen atom; iso- refers to the occurrence of the hydrocarbon fragment $(CH_3)_2CH—$ in the structure; and *tertiary-* or *tert-* indicates attachment to the substituent group to a carbon atom carrying no hydrogen atoms. The terms are widely used in organic nomenclature. Systematic methods of naming organic compounds are also in use: e.g. if the longest single carbon chain in an aliphatic compound is numbered, the position of substituent groups can readily be indicated. Examples of such a system are 1-butene and 2-bromobutane (p. 278). Isobutyl chloride is alternatively named 1-chloro-2-methylpropane; *tert*-butyl chloride becomes 2-chloro-2-methylpropane.

19.8. Raw material sources

Hydrocarbon types in the main form the source materials for industry. (This does not infer that an organic compound will always be prepared directly from a hydrocarbon; the synthetic methods employed both in the laboratory and in industry are many and varied. Hydrocarbon structures are often, but not always, suitable starting points.) Raw materials are obtained from two principal sources.

Coal is by far the largest known indigenous source in the U.K., and on that account should be regarded as a material of the highest importance. Extensive deposits occur elsewhere in Europe, in the North American continent, and in many other places. Currently world reserves are estimated to be of the order of 8×10^{12} tons. Coal is not a single substance but a complex mixture of large molecules, mainly aromatic hydrocarbon in type. While employed chiefly as a fuel, it has in the past been a rich source of aromatic compounds. Less conveniently aliphatic types can also be obtained from coal.

Petroleum is also an organic material, but is principally aliphatic and alicyclic in character. Again regarded first as a fuel, it has now far outstripped coal as the prime source of organic raw materials. Essential aliphatics can readily be obtained from it, and secondary processes are available for converting these to aromatics. The single disadvantage of petroleum in the U.K. and Western Europe is the heavy dependence that must be placed on imported supplies, at least for the present.

The first product of a petroleum well is not oil, but a mixture, rich in methane and other hydrocarbon gases, called *natural gas*. Natural gas is sometimes unassociated with the higher boiling liquid petroleum products and can well be regarded as a raw material in its own right. It is a very valuable product, both as a fuel and as a source material. Large deposits of natural gas discovered in the North Sea off the coast of Holland prompted a wide search for gas off the east coast of U.K. Successful strikes were first made in 1965, and these have since multiplied at a spectacular rate. The long-term effect of natural gas on the U.K.'s fuel policy, and on the organic chemical industry, is likely to be a major one. Elsewhere in Europe, France and Italy have been notably successful in making rich natural gas strikes. Abundant supplies are available in the U.S.A., where it is derived from both gas and petroleum wells.

CHAPTER 20

Hydrocarbons: I. Aliphatic Compounds

ALIPHATIC hydrocarbons are divided into three families: alkanes (or paraffins), alkenes (or olefins) and alkynes (or acetylenes). Alkanes are abundantly available from natural sources: this is not the case with the industrially important alkenes and alkynes, which have to be obtained by synthesis.

Alkanes

20.1. Alkanes from petroleum

Crude petroleum is a highly complex hydrocarbon mixture. Alkanes and alicyclics (or naphthenes) predominate, and aromatics are present in varying minor quantities. Alkanes are obtained from petroleum by fractional distillation, those of particular use to the chemical industry being the early members which up to the C_4 fraction exist as gases. Liquid petroleum contains a large quantity of these gases, perhaps as much as 200 times its own volume.

Natural gas is also a rich source of the lower alkanes, particularly methane, which is always by far the largest constituent. Early reports of British North Sea gas suggest that it contains upward of 90 per cent methane.

Relatively pure single components can be obtained from natural gas or petroleum by extensive refining. Thus propane and butane, liquefied and stored under pressure, are available to the commercial and domestic market. (These products are also obtained from the breakdown of higher petroleum fractions.) The liquid fractions from petroleum, however, are not normally separated into pure components. The early liquid fractions are widely marketed as solvents, and for this purpose a hydrocarbon mixture is quite suitable. Hence '40–60' petroleum ether is a hydrocarbon mixture distilling between 40°C and 60°C, and is essentially the C_5–C_6 fraction. Two '40–60' petroleum ethers may differ greatly in their detailed chemical analysis because of differences in source, but their solvent powers will be very similar.

20.2. Reactions of alkanes

Alkanes are not very reactive. Being saturated and having no labile functional group, they are chemically stable, at least under mild conditions. Some of their more important reactions include:

(a) *Substitution reactions*

The substitution of hydrogen by other elements in an alkane can occur progressively. Hydrocarbon chlorination illustrates this reaction.

$$CH_4 \xrightarrow{Cl_2} CH_3Cl \xrightarrow{Cl_2} CH_2Cl_2 \xrightarrow{Cl_2} CHCl_3 \xrightarrow{Cl_2} CCl_4$$

| Methane | Methyl chloride | Methylene chloride | Chloroform | Carbon tetrachloride |

(b) *Cracking reactions*

The decomposition of a raw material often leads to useful products. If the decomposition is induced by high temperature, of the order of 400°C and above, the reaction is described as *pyrolysis*, or *carbonization*, or *cracking*. This last expression is the one most commonly used in referring to the thermal decomposition of alkanes as practised in the petrochemical industry.

When alkanes are cracked, multiple reactions occur, and reaction variables such as temperature, catalyst, and contact time have to be optimized for a given result. Three types of reaction can be distinguished.

(i) Alkenes, outstandingly important starting materials for further synthesis, are produced along with hydrogen. Thus ethane cracking yields ethylene.

$$CH_3 \cdot CH_3 \longrightarrow CH_2 = CH_2 + H_2$$
$$\text{Ethane} \qquad \text{Ethylene}$$

(ii) Smaller molecules are produced by breaking carbon–carbon bonds. If this occurs, alkenes again appear in the cracked products.

$$CH_3 \cdot CH_2 \cdot CH_3 \longrightarrow CH_2 = CH_2 + CH_4$$
$$\text{Propane} \qquad \text{Ethylene} \quad \text{methane}$$

(iii) Rearrangement, or isomerization, takes place to give, for example, branched chain isomers from *n*-alkanes.

$$CH_3 \cdot CH_2 \cdot CH_2 \cdot CH_3 \longrightarrow \underset{\text{Isobutane}}{CH_3 \cdot \overset{\overset{\displaystyle CH_3}{\displaystyle |}}{CH} \cdot CH_3}$$
$$\underset{n\text{-Butane}}{}$$

Extensive industrial use is made of these reactions, all of which tend to occur in an increasingly complex fashion as the average chain length of the alkane feed to the cracker grows. In any high-temperature pyrolysis elemental carbon is also produced in quantities which increase with the severity of the reactions conditions.

(c) *Combustion*

When an alkane reacts with oxygen, the end products are carbon dioxide, water, and a large quantity of energy, most of which is released as heat. Thus the reaction between methane and oxygen is fully represented by the equation:

$$CH_4 + 2O_2 CO_2 + 2H_2O(l) \Delta H = -212 \text{ kcal}$$

The exothermic reaction of hydrocarbons with oxygen is

made use of in a wide range of gaseous and liquid fuels derived from petroleum. The heat produced may be used directly, or it may be converted to a power stroke as in the internal combustion engine.

(d) *Controlled oxidation*

Oxidation of alkanes need not be carried to completion. If the reaction is carried out under controlled conditions, a variety of important oxygen-containing compounds is obtained. As yet no extensive application has been made of reactions of this sort, for too often even so-called 'controlled' oxidation yields product mixtures too complex to be useful.

Alkenes

Alkenes constitute an homologous series C_nH_{2n}. The early members are listed in Table 20.1. The outstanding structural feature common to all alkenes, the functional group of the series in fact, is the carbon–carbon double bond.

Alkenes have in high degree a chemical versatility which has helped to make them a corner stone of the modern chemical industry. The range of materials now produced in whole or in part from the simple alkenes up to the C_4 members is quite remarkable.

TABLE 20.1. ALKENES

Name	Structure
Ethylene	$CH_2{=}CH_2$
Propylene	$CH_2{=}CH{\cdot}CH_3$
1-Butene	$CH_2{=}CH{\cdot}CH_2{\cdot}CH_3$
2-Butene	$CH_3{\cdot}CH{=}CH{\cdot}CH_3$
Isobutene (2-methylpropene)	$CH_2{=}C{\cdot}CH_3$ $\quad\quad\ \ \vert$ $\quad\quad\ \ CH_3$
1-Pentene	$CH_2{=}CH{\cdot}CH_2{\cdot}CH_2{\cdot}CH_3$

20.3 Synthesis of alkenes

(a) *Cracking of alkanes*

The important lower alkenes are produced by alkane cracking. If abundant supplies of natural gas are available, as in the U.S.A., this is used as feedstock. In Europe, the naphtha fraction from petroleum is used, although this practice may well alter as natural gas finds multiply.

Lower alkenes can be obtained as pure components by fractional distillation of cracked vapours. *Steam cracking* is widely employed in a number of processes where ethylene production is the main aim. In this technique, the introduction of superheated steam into the cracking chamber decreases losses due to elemental carbon formation, and increases ethylene yield by permitting operation at a higher temperature than would otherwise be possible — cracking temperatures of about 900°C have been reported.

(b) *Dehydration of alcohols*

The functional group of the alcohol family is the hydroxyl — OH group. If a molecule of water is removed from an alcohol, an alkene results.

With petroleum sources increasingly available, alcohol dehydration on a large scale has diminished in importance as a source of alkenes. Where practised, the alcohol is vaporized and passed over a suitable dehydrating catalyst.

$$CH_3 \cdot CH_2OH \xrightarrow[\text{Alumina}]{300°C} CH_2 = CH_2 + H_2O$$

Ethyl alcohol   Ethylene

20.4. Reactions of alkenes

Alkenes, being unsaturated, are very reactive and characteristically undergo a wide range of *addition reactions*, several of which are of interest.

(a) *Hydrogenation*

e.g.

$$CH_3 \cdot \underset{\underset{CH_3}{|}}{\overset{\overset{CH_3}{|}}{C}} \cdot CH = \overset{\overset{CH_3}{|}}{C} \cdot CH_3 + H_2 \xrightarrow[200°C/5-10 \text{ atm}]{\text{Ni catalyst}} CH_3 \cdot \underset{\underset{CH_3}{|}}{\overset{\overset{CH_3}{|}}{C}} \cdot CH_2 \cdot \overset{\overset{CH_3}{|}}{CH} \cdot CH_3$$

2,4,4-Trimethyl-2-pentene   2,2,4-Trimethylpentane

(b) *Halogen addition*

e.g.

$$CH_3 \cdot CH = CH_2 + Cl_2 \longrightarrow CH_3 \cdot \overset{\overset{Cl}{|}}{CH} \cdot CH_2Cl$$

Propylene   1,2-Dichloropropane

(c) *Halogen acid addition*

e.g.

$$CH_2 = CH_2 + HBr \longrightarrow CH_3 \cdot CH_2Br$$

Ethylene   Ethyl bromide

This reaction has limitations. If the alkene has a nonsymmetrical structure, isomers are produced:

$$CH_3 \cdot CH_2 \cdot CH = CH_2 + HCl \begin{cases} \rightarrow CH_3 \cdot CH_2 \cdot \overset{\overset{Cl}{|}}{CH} \cdot CH_3 \\ \quad\; Sec\text{-Butyl chloride} \\ \\ \rightarrow CH_3 \cdot CH_2 \cdot CH_2 \cdot CH_2Cl \\ \quad\; n\text{-Butyl chloride} \end{cases}$$

1-Butene

In such a case, the major reaction product can be predicted by applying an empirical rule which states that the *hydrogen of the acid adds to the carbon atom already the richer in hydrogen*. In the above example, the chief product would therefore be *sec*-butyl chloride.

(d) *Sulphuric acid addition: alkene hydration*

The reactions of sulphuric acid, H_2SO_4, in organic chemistry are better understood in the light of a more explicit structural formula, $HO \cdot SO_2 \cdot OH$. It will add across a double bond as the acid HX, X being the group $-O \cdot SO_2 \cdot OH$,

e.g. $CH_2{=}CH_2 + HO \cdot SO_2 \cdot OH \longrightarrow CH_3 \cdot CH_2 \cdot O \cdot SO_2 \cdot OH$
Ethylene   Ethyl hydrogen sulphate

Subsequent hydrolysis of the addition compound yields an alcohol and regenerates sulphuric acid:

$$CH_3 \cdot CH_2 \cdot O \cdot SO_2 \cdot OH + H \cdot OH \longrightarrow CH_3 \cdot CH_2 \cdot OH + H_2SO_4$$
Ethyl alcohol

This is a very important method of synthesizing alcohols. We may consider the reaction to be the addition of water (hydration) to a double bond under the influence of acid catalyst. Whether it be water, $H \cdot OH$, which adds directly, or sulphuric acid, $HO \cdot SO_2 \cdot OH$, which first reacts, the direction of addition is similar to that of the halogen acids. Hence propylene yields isopropyl alcohol, not *n*-propyl alcohol,

$$CH_3 \cdot CH{=}CH_2 + H \cdot OH \xrightarrow{\text{H}^+} CH_3 \cdot \overset{\displaystyle OH}{\overset{|}{CH}} \cdot CH_3$$
Propylene   Isopropyl alcohol

(e) *Alkylation*

Alkylation is the term used to describe the addition of an alkane to an alkene. The general reaction may be represented by:

$$>C{=}C< + RH \rightarrow >CH{-}\underset{\displaystyle R}{\overset{\displaystyle |}{C}}<$$

Alkylation is a key step in a number of petroleum-based processes. Alkanes used in gasoline production can be formed by this method:

e.g.

$$CH_3 \diagdown C{=}CH_2 + HC{-}CH_3 \longrightarrow CH{\cdot}CH_2{\cdot}C{\cdot}CH_3$$

CH₃\
 C=CH₂ + HC—CH₃ ⟶   CH·CH₂·C·CH₃\
CH₃   CH₃   CH₃   CH₃

Isobutene	Isobutane	Iso-octane (2,2,4-Trimethylpentane)

If in place of the alkane aromatic hydrocarbons are used important aromatic derivatives can be obtained.

(f) *Polymerization*

Polymerization is the process whereby a large number of small molecules are linked together chemically to form a final product of very high molecular weight. The starting material is called *monomer*, the end product *polymer*. Polymers are very important, and a later chapter (Chapter 27) is devoted to them.

Alkenes can be induced to undergo polymerization. We may picture the initial reaction to be one between some highly reactive chemical fragment $B-$ and an alkene molecule $CH_2{=}CHX$:

$$B{-} + CH_2{=}CH \longrightarrow B{-}CH_2{\cdot}CH{-}$$
$$\quad\quad\quad\quad X \quad\quad\quad\quad X$$

Initiation is followed by successive addition of monomer molecules to the reactive site, until a polymer chain containing perhaps thousands of monomer units is obtained:

$$B{\cdot}CH_2{\cdot}CH{-} + n{\cdot}CH_2{=}CH \longrightarrow B{\cdot}CH_2{\cdot}CH{\cdot}(CH_2{\cdot}CH)_n$$
$$\quad X \quad\quad\quad\quad\quad X \quad\quad\quad\quad\quad X \quad\quad\quad\quad X$$

The reaction is terminated by one of a number of possible side reactions.

20.5. Dienes

Dienes are compounds containing two carbon–carbon double bonds. Much of their chemistry is similar to that of simple alkenes, i.e. typically they take part in addition reactions involving the unsaturated bonds.

Three important members of this group are shown below:

$$CH_2{=}CH{\cdot}CH{=}CH_2 \qquad CH_2{=}\underset{\substack{| \\ CH_3}}{C}{\cdot}CH{=}CH_2$$

Butadiene   Isoprene

$$CH_2{=}\underset{\substack{| \\ Cl}}{C}{\cdot}CH{=}CH_2$$

Chloroprene

These three dienes form the basis of the modern synthetic rubber industry (Chapter 27). Butadiene and isoprene can be produced by cracking of suitable C_4 and C_5 petroleum fractions; other methods are also available and in use. Chloroprene may be synthesized from acetylene (Section 20.7).

Alkynes

Alkynes, or acetylenes, constitute the homologous series C_nH_{2n-2}. The recurring functional group in the alkynes is a carbon–carbon triple bond. The series is typified by the parent member acetylene $HC{\equiv}CH$, the only alkyne of any industrial significance.

20.6. Synthesis of acetylene

(a) *The carbide process.*

When carbon is fused with calcium oxide in an electric furnace at about 2500–3000°C, calcium carbide is formed. This in turn decomposes in water to form acetylene.

$$CaO + 3C \longrightarrow Ca_2C_2 + CO$$
Calcium
carbide

$$Ca_2C_2 + 2H_2O \longrightarrow C_2H_2 + Ca(OH)_2$$
Acetylene

Coke is used as the carbon source, limestone as the calcium oxide source. The ready availability of these materials and the comparative simplicity of the process has made this the premier method of industrial acetylene production in the past.

(b) *Petroleum sources*

Like alkenes, acetylene can be obtained from alkane feedstock. Methane is cracked at temperatures of about 1500°C to give acetylene and hydrogen:

$$2CH_4 \longrightarrow C_2H_2 + 3H_2$$

The high temperatures required can be achieved in a furnace, but electric arc cracking is preferred in some processes.

Methane cracking is superseding the carbide route in acetylene manufacture, particularly in areas where cheap supplies of natural gas are available. Higher hydrocarbon fractions such as naphtha can be cracked to acetylene, but they are less suitable than methane.

20.7.   Reactions of acetylene

Being highly unsaturated acetylene undergoes a wide range of addition reactions, and much of its chemistry bears comparison with that of the alkenes. The triple bond in acetylene increases its potentiality in industrial synthesis, so much so that acetylene may be classified as a raw material in its own right, distinct from both coal and petroleum. Some of the more important reactions are summarized below:

(a) *Halogenation*

e.g. $CH{\equiv}CH + 2Cl_2 \longrightarrow ClCH{=}CHCl + Cl_2 \longrightarrow$
Dichloroethylene

$$Cl_2CH \cdot CHCl_2$$
Tetrachloroethane

(b) *Addition of HX compounds*

e.g.
$$CH{\equiv}CH + HCl \longrightarrow CH_2{=}CH{\cdot}Cl$$
Vinyl chloride

$$CH{\equiv}CH + HCN \longrightarrow CH_2{=}CH{\cdot}CN$$
Acrylonitrile

$$CH{\equiv}CH + HO_2C{\cdot}CH_3 \longrightarrow CH_3{=}CH{\cdot}O_2C{\cdot}CH_3$$
Acetic acid   Vinyl acetate

Each of these reactions yields a vinyl compound capable of polymerizing to important products.

As with alkenes, the hydration of acetylene may be considered as an 'HX-type' addition. The initial product, however, is unstable, but the spontaneous rearrangement which occurs provides a route (of decreasing importance) to acetaldehyde:

$$CH{=}CH + H_2O \rightarrow [CH_2{=}CH{\cdot}OH] \rightarrow CH_3{\cdot}CHO$$
'Vinyl alcohol' Acetaldehyde

(c) *Dimerization*

Polymerization was defined as the linking together of many molecules. *Dimerization* is the linking together of two molecules, and applied to acetylene gives a route to the diene chloroprene.

$$2CH{\equiv}CH \rightarrow CH_2{=}CHC{\equiv}CH \xrightarrow{HCl} CH_2{=}CH{\cdot}\underset{|}{\overset{C_1}{C}}{=}CH_2$$
Vinylacetylene   Chloroprene

(d) *Combustion*

If acetylene is burned in oxygen, a flame temperature of about 2300°C can be obtained, a temperature which is substantially higher than that given by any other common gaseous fuel. This feature is made use of in the oxyacetylene burner commonly used in welding and metal-cutting operations.

CHAPTER 21

Hydrocarbons: II. Aromatic Compounds

EARLY in the development of organic chemistry there emerged a group of compounds clearly related to one another, and distinctly different from the aliphatics. They generally had pleasant odours, so were called *aromatic*. Michael Faraday is credited with the first isolation in 1825 of the parent compound of the aromatic class, benzene.

21.1. The structure of benzene

Benzene is a hydrocarbon; its molecular formula is C_6H_6. On the basis of the hydrogen:carbon ratio we would expect it to display a high degree of unsaturation:

		$H:C$ ratio	Characteristics
Methane	CH_4	4:1	saturated
Ethylene	C_2H_4	2:1	unsaturated
Acetylene	C_2H_2	1:1	highly unsaturated
Benzene	C_6H_6	1:1	?

Benzene can under certain conditions undergo addition reactions of the type associated with multiple carbon–carbon bonds of the alkene or alkyne class. This behaviour, however, is not typical of benzene and its derivatives. Aromatic compounds more characteristically undergo substitution reactions. The behaviour of benzene, then, is that of a saturated hydrocarbon despite its hydrogen:carbon ratio.

The apparent anomalies in benzene chemistry pose problems in structure formulation which have been reconciled by postulating the ring structure shown below:

H—C—C—H
H—C—C—H
H

Benzene

The lack of alkene double bond character is due to delocalization of the *p*-orbital electrons which form the second bond of any carbon–carbon double bond (Section 1.8). The structure of benzene, i.e. a planar arrangement of carbon and hydrogen atoms with symmetrical delocalized electron clouds above and below the plane of the ring, is best represented by the symbol:

Benzene: a symmetrical structure

Molecular orbital theory has led to a definition of aromaticity, or aromatic character, much wider than once was the case. For our purposes, however, we shall remain conservative and define an aromatic compound as being a compound containing at least one benzene-type ring. Compounds so defined are sometimes called *benzenoid compounds*. Although the chemical behaviour exhibited by the ring in any one benzenoid compound is modified in some way by substituted side groups, the reactions characteristic of benzene are characteristic of the aromatic ring in any compound of this class.

21.2. Synthesis of aromatic hydrocarbons

(a) *Coal sources*

When coal is carbonized it decomposes into four primary fractions: coal gas, ammoniacal liquor, coke and coal tar. This last fraction, which probably accounts for less than 10 per cent of the coal charge, is rich in essential aromatic compounds which can be isolated by distillation and further refining. The more important hydrocarbon constituents obtained from coal tar sources are listed in Table 21.1. Their detailed structures we shall consider later.

TABLE 21.1. ESSENTIAL
HYDROCARBONS FROM
COAL TAR

Product	Molecular formula
Benzene	C_6H_6
Toluene	C_7H_8
Xylene	C_8H_{10}
Naphthalene	$C_{10}H_8$
Anthracene	$C_{10}H_{14}$
Phenanthrene	$C_{14}H_{10}$

The first fraction collected from coal tar is called *benzol*, essentially a benzene-toluene mixture containing in addition a little xylene. Motor benzol finds good use as a high-octane gasoline fuel. (Crude benzol actually separates in the course of the primary coal carbonization, and is often considered to be a fraction in its own right.)

(b) *Petroleum sources*

In the U.S.A. upward of 75 per cent of aromatic hydrocarbons are derived from petroleum. In Western Europe coal is still the principal source of aromatics, but the challenge from petro-chemical sources is growing.

A number of routes are used to obtain aromatic hydrocarbons from petroleum. Some crude oils contain a minor but significant percentage of aromatics which can be extracted directly. This is not common, and synthetic techniques usually have to be adopted.

Aromatization of other hydrocarbon types can be applied

most obviously to cycloalkanes, the cycloalkane structure defining the aromatic compound obtained:

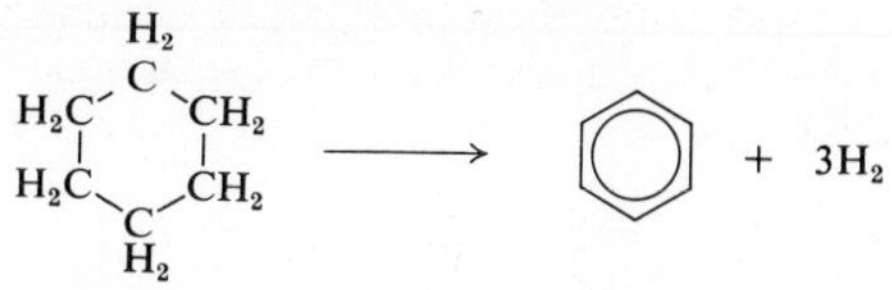

Cyclohexane   Benzene

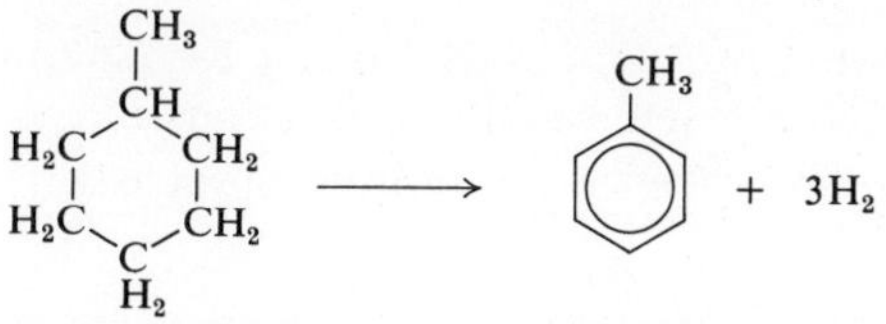

Methylcyclohexane   Toluene

Conversion to aromatic structures is not confined to cyclo-alkanes, however. Alkanes can also undergo cyclization and aromatization, the alkane structure again determining the aromatic compound or mixture obtained:

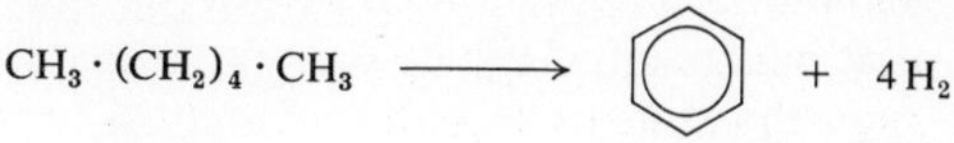

$$CH_3 \cdot (CH_2)_4 \cdot CH_3 \longrightarrow \text{⬡} + 4H_2$$

n-Hexane   Benzene

$$CH_3 \cdot (CH_2)_6 \cdot CH_3 \longrightarrow \text{Xylenes} + \text{Ethylbenzene} + 4H_2$$

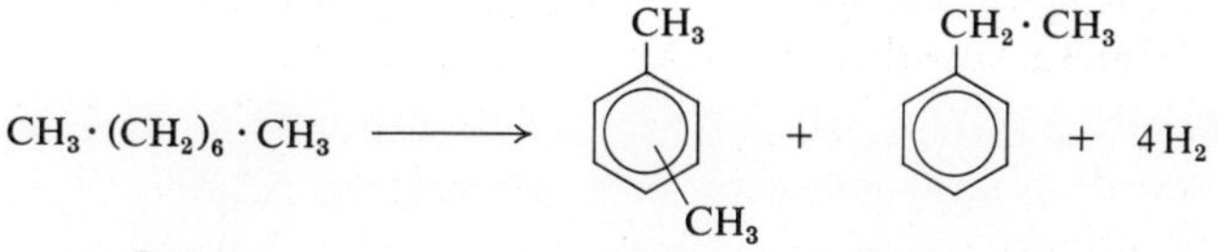

n-Octane   Xylenes   Ethylbenzene

The above operation, whether applied to cycloalkanes or alkanes, is called hydroforming, or platforming if a platinum-containing catalyst is used. The preferred feedstock is a light petroleum fraction rich in suitable cycloalkanes.

A second and distinctly different technique has been applied to alkane-rich fractions whereby cracking is used first to produce alkenes and dienes which then interact to form aromatics:

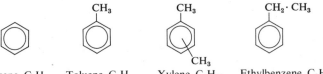

Butadiene   Ethylene   Cyclohexene   Benzene

Replacement of ethylene by propylene in the above reaction would yield toluene.

Aromatization procedures lead to the formation of complex mixtures which are separated into fractions of required specification by a number of further refining processes.

21.3.   Homologues of benzene

Some of the more important aromatic hydrocarbons together with their molecular formulae have been listed in Table 21.1. There is no obvious pattern within these formulae that we would readily recognize as representing an homologous series similar to those of the aliphatic hydrocarbons. How then does homology occur in aromatic compounds? The answer to this question is that benzene has the distinction of being the parent member of more than one hydrocarbon family. The compounds indicated in Table 21.1 belong to one or other of the two major aromatic hydrocarbon series, the first of which results from alkyl radical substitution in the benzene ring:

Benzene, C_6H_6 Toluene, C_7H_8 Xylene, C_8H_{10} Ethylbenzene, C_8H_{10}

The above series, to which is given the name *arenes*, is a mixed hydrocarbon family in that each member is part aromatic, part aliphatic; the presence of the benzenoid ring, however, has such a dominating effect on the chemistry of the arenes that they are usually regarded as aromatic.

Aromatic hydrocarbon structures can also be obtained by the fusing together of two or more benzene rings, as in naphthalene, $C_{10}H_8$:

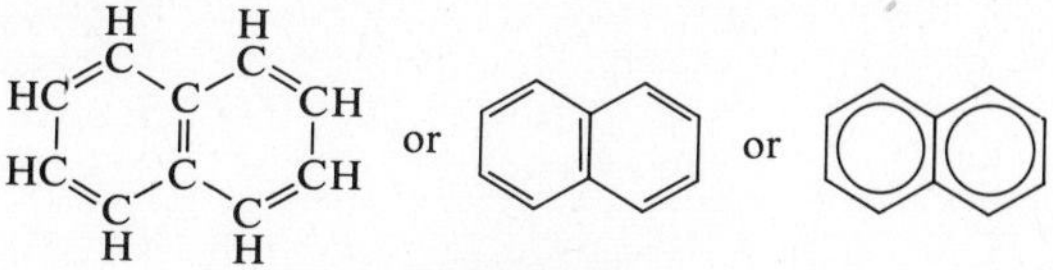

Naphthalene

In naphthalene all carbon atoms are no longer identical; nor indeed are the eight hydrogen atoms, either in position or in relative reactivity. Naphthalene is the second member of the series of *fused polynuclear hydrocarbons*. Two isomers, anthracene and phenanthrene, constitute the next member of this series, $C_{14}H_{10}$.

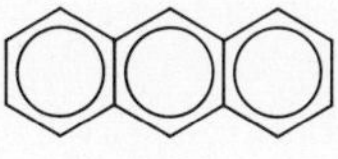

Anthracene 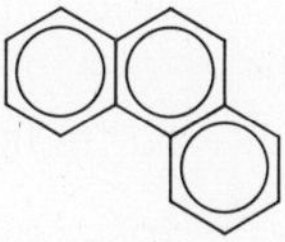   Phenanthrene

and multiple ring systems are derived by permutations on this type of ring extension. A number of the larger structures have received much attention as possible cancer-inducing compounds (carcinogens). They arise in several high temperature pyrolytic or combustion systems. Tobacco smoke is currently the most notorious source of hydrocarbon carcinogens; active constituents have also been reported in, for example, coal tar and soot.

21.4. Isomerism and nomenclature in benzene derivatives

If the benzene ring is monosubstituted by some group X, the resulting structure, viz.

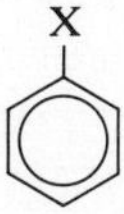

or C_6H_5X or Ph·X

is quite unambiguous. No isomerism is possible, for in the completely symmetrical benzene ring all the positions are identical. Some important monosubstituted benzenes are given in Table 21.2.

TABLE 21.2. MONOSUBSTITUTED
DERIVATIVES OF BENZENE

Name	Formula
Nitrobenzene	$C_6H_5 \cdot NO_2$
Aniline	$C_6H_5 \cdot NH_2$
Chlorobenzene	$C_6H_5 \cdot Cl$
Phenol	$C_6H_5 \cdot OH$
Benzene sulphonic acid	$C_6H_5 \cdot SO_3H$
Benzoic acid	$C_6H_5 \cdot CO_2H$

Monosubstitution in benzene, however, results in a non-symmetrical molecule in which *three* different positions may be discerned. If we number round the ring from the principal functional group,

numbering in benzene rings

then positions 2 and 6, called the *ortho positions*, will be seen to be equivalent; positions 3 and 5, called the *meta positions*, will be seen to be equivalent; and the remaining position 4, called the *para position*, to be unique in itself. On this basis, disubstitution should yield three isomers. This argument is well proven, and xylene provides us with a convenient example:

Ortho- or *o*-Xylene *Meta-* or *m*-Xylene *Para-* or *p*-Xylene

If more than two groups are substituted in the ring, the number of possible isomers may or may not increase; thus there can only be one hexamethylbenzene, while there are three trimethylbenzenes.

21.5. Reactions of aromatic hydrocarbons

While reactions in a particular aromatic nucleus are affected by groups substituted in the nucleus, there is a clear pattern of behaviour which permits us to consider the reactions of benzene to be as typical of those of any aromatic compound as, say, the reactions of ethylene are of any alkene. The difference between benzene and its many relations can be regarded for our purposes as one of degree rather than of type. If the chemical reactivity of benzene is taken as the norm for aromatic compounds, other aromatic rings are found by comparison to exhibit either enhanced or suppressed reactivity, and reaction conditions have to be adjusted accordingly; but the same type of chemical change can usually be brought about.

Substitution is the normal mode of reaction in aromatic nuclei. Two addition reactions in benzene which are important are described below; these, however, must be regarded as abnormal.

(a) *Substitution reactions*

Substitution in the aromatic nucleus is an extensively used reaction in industry. Some more important substitution reactions are shown below:

(i) *Alkylation*

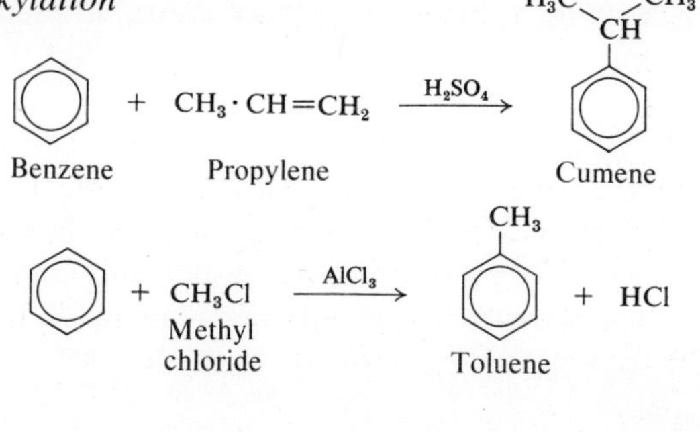

Benzene + CH₃·CH=CH₂ (Propylene) —H₂SO₄→ Cumene

Benzene + CH₃Cl (Methyl chloride) —AlCl₃→ Toluene + HCl

Alkylating reagents other than alkenes or alkyl halides can be used.

(ii) *Nitration.* The nitro-NO_2 group is substituted in the benzene ring by reaction with concentrated 'mixed acid', a mixture of nitric and sulphuric acids. Unless the temperature is held at a moderate level, disubstitution may occur:

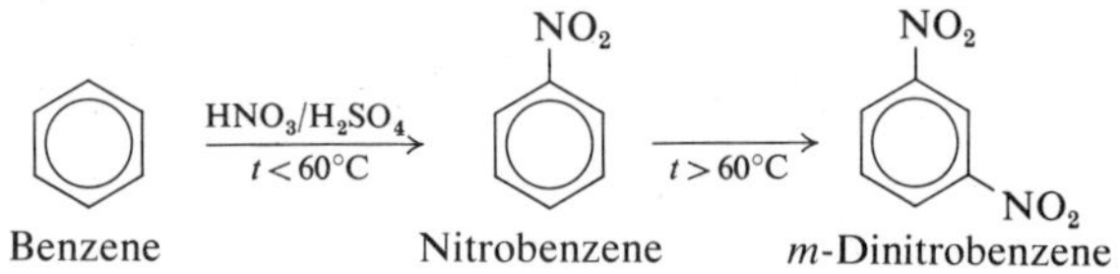

Benzene $\xrightarrow[t < 60°C]{HNO_3/H_2SO_4}$ Nitrobenzene $\xrightarrow{t > 60°C}$ *m*-Dinitrobenzene

(iii) *Sulphonation*

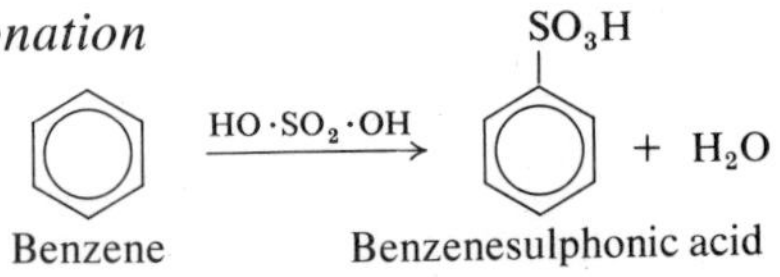

Benzene $\xrightarrow{HO \cdot SO_2 \cdot OH}$ Benzenesulphonic acid $+ \ H_2O$

Sulphonation is much used in dyestuff, drug and other industries to confer water solubility on organic products.

(iv) *Halogenation*

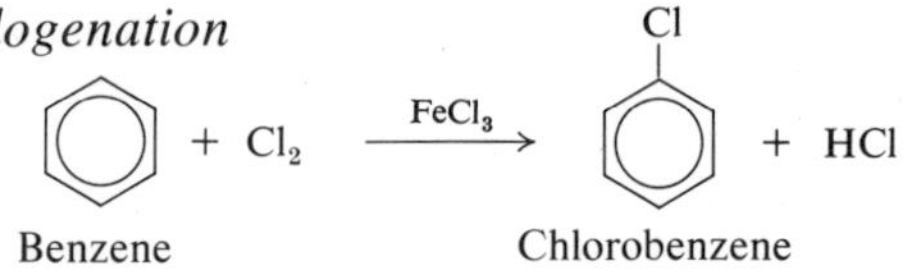

Benzene $+ \ Cl_2$ $\xrightarrow{FeCl_3}$ Chlorobenzene $+ \ HCl$

(b) *Addition reactions*

Benzene and chlorine undergo an addition reaction in the presence of ultraviolet-light forming benzene hexachloride, $C_6H_6Cl_6$, one form of which is a potent insecticide, Gammexane.

Benzene and hydrogen also take part in an addition reaction

if a suitable catalyst is present, reversing the hydroforming reaction:

Benzene $+ 3H_2 \longrightarrow$ Cyclohexane

21.6. Polysubstitution in the benzene ring

We have seen that the introduction of a second group into the benzene ring can yield three isomers. An empirical rule permits us to predict the *orientation* of a second substituent very easily. It states that if a compound C_6H_5X is to be disubstituted, *the orientation of the second group is dependent solely*

TABLE 21.3. PRINCIPAL SUBSTITUENT GROUPS IN AROMATIC RINGS

o and *p*-directing	*m*-directing
—R (alkyl benzenes)	—NO₂ (nitro compounds)
—OH (phenols)	—SO₂—OH (sulphonic acids)
—NH₂ (amines)	—C(=O)—OH (carboxylic acids)
—Halogen (halogen compounds)	—C(=O)—R (aldehydes and ketones)

on the nature of the group X; if X is a saturated group, a mixture of o- *and* p-*isomers results, if unsaturated the* m-*isomer results.*

There are exceptions to this rule, but they are comparatively rare. The structures of the principal substituent groups and their directing influences are indicated in Table 21.3. Substitution beyond the second stage will be affected by the directing influence of each group already in the ring, i.e. the groups X and Y. There is no simple rule to predict the net effect.

CHAPTER 22

Hydroxy Compounds

COMPOUNDS in which the hydroxyl—OH group is the principal functional group may be divided into two classes, *alcohols* and *phenols*. In the former class the hydroxyl function is bonded to an aliphatic carbon atom, in the latter to an aromatic carbon atom:

$$C_{al}—OH \qquad C_{ar}—OH$$
Alcohol \qquad\qquad Phenol

There are similarities between the behaviour of alcohols and that of phenols but there are also striking differences, and we shall consider the two classes separately.

Alcohols

22.1. Structure and nomenclature

Alcohols are the key intermediates in many syntheses, and by virtue of their physical properties form a valuable group of industrial and commercial solvents. In part their usefulness as solvents stems from the freely water-soluble properties displayed by the lower members. This affinity for aqueous systems is not surprising in the light of the structural similarities between water, $H \cdot OH$, and alcohols, $R \cdot OH$. As the organic part of the molecule increases in size, so the water solubility decreases. The comparison of alcohols with water can be extended; e.g. both exhibit hydrogen bonding (Section 1.9).

Some common alcohols are listed in Table 22.1 where water solubilities are also given. Use of systematic names (methanol, etc.) for the lower alcohols is widespread.

TABLE 22.1. ALCOHOLS

Name	Structure	Solubility in water (g/100 g)
Methyl alcohol (Methanol)	$CH_3 \cdot OH$	Soluble in all proportions
Ethyl alcohol (Ethanol)	$CH_3 \cdot CH_2 \cdot OH$	Soluble in all proportions
n-Propyl alcohol	$CH_3 \cdot CH_2 \cdot CH_2 \cdot OH$	Soluble in all proportions
Isopropyl alcohol	$CH_3 \cdot \underset{\underset{OH}{\mid}}{CH} \cdot CH_3$	Soluble in all proportions
n-Butyl alcohol	$CH_3 \cdot (CH_2)_3 \cdot OH$	7·9
Isobutyl alcohol	$CH_3 \cdot \underset{\underset{CH_3}{\mid}}{CH} \cdot CH_2OH$	10·0
sec-Butyl alcohol	$CH_3 \cdot \underset{\underset{OH}{\mid}}{CH} \cdot CH_2 \cdot CH_3$	12·5
tert-Butyl alcohol	$(CH_3)_3 \cdot C \cdot OH$	Soluble in all proportions
Benzyl alcohol	$C_6H_5 \cdot CH_2OH$	4

22.2. Synthesis of alcohols

On the industrial scale alcohols are usually obtained from alkenes, either by the hydration reaction (Section 20.4) or by the Oxo process. In the latter process carbon monoxide–hydrogen mixtures (known generally as *synthesis gas*) are reacted with alkenes to give mixtures of aldehydes which can then be catalytically hydrogenated to alcohols:

$$R \cdot CH {=\!=} CH_2 + CO + H_2$$

$$\rightarrow \underset{\text{Aldehyde}}{R \cdot CH_2 \cdot CH_2 \cdot CHO} \xrightarrow{H_2/Ni} \underset{\text{Alcohol}}{R \cdot CH_2 \cdot CH_2 \cdot CH_2OH}$$

$$\rightarrow \underset{\text{Aldehyde}}{R \cdot \underset{\underset{CHO}{\mid}}{CH} \cdot CH_3} \xrightarrow{H_2/Ni} \underset{\text{Alcohol}}{R \cdot \underset{\underset{CH_2OH}{\mid}}{CH} \cdot CH_3}$$

The reaction is often referred to as *hydroformylation* to indicate the addition of a hydrogen atom H and a formyl group —CHO across the double bond.

Fermentation can also be used to obtain alcohols. Fermentation is the degradation of naturally occurring substances into mixtures of simpler units brought about by the action of highly specific biological catalysts called enzymes. Alcohols are among the products of carbohydrate fermentation. Early man was quick to recognize this, and fermentation processes as practised in the brewing and distilling industries have a long history. Prior to the emergence in the last 40 years or so of comprehensive petroleum-based methods, however, fermentation was also used to produce a number of aliphatic organic chemicals for general industrial use. These included alcohols, ethyl alcohol usually being the major component, while others up to the C_5 members were produced in commercially useful quantities. Despite competition from other sources, fermentation remains an important route to industrial alcohols.

Fermentation does not result in the production of methyl alcohol, which is highly toxic and must never be confused with its near neighbour ethyl alcohol. Methyl alcohol is also unobtainable from the other alkene routes. It is made from synthesis gas of appropriate composition in a catalysed high-pressure process:

$$CO + 2H_2 \longrightarrow CH_3OH$$
Methyl alcohol

For many purposes, alcohol mixtures are often quite suitable. Ethyl alcohol solvent is usually produced as an azeotropic mixture of about 96 per cent alcohol and 4 per cent water. *Methylated spirits* is a name applied to a number of mixtures, usually rich in ethyl alcohol and containing a little methyl alcohol and other constituents. Essentially pure ethyl alcohol is referred to as *absolute alcohol*.

22.3. Reactions of alcohols

Alcohols can readily take part in a wide variety of reactions. They are ideal starting materials for the synthesis of most of the more important aliphatic compounds, as indicated below:

(a) *Dehydration*

The dehydration of an alcohol to produce an alkene (Section 20.3) is less useful in industrial practice than is the reverse hydration reaction now that alkenes are so freely available from petroleum.

(b) *Alkyl halide formation*

The hydroxyl group in alcohols can be replaced by the halogen atom of a hydrogen halide.

$$R \cdot OH + HX \rightarrow R \cdot X + H_2O \qquad X = \text{halogen}$$

(c) *Ether formation*

Ethers form a family of compounds of general structure $R \cdot O \cdot R'$. Their principal use is as solvents. The best known, diethyl ether, $CH_3 \cdot CH_2 \cdot O \cdot CH_2 \cdot CH_3$, is also used as an anaesthetic.

Alcohols can form ethers according to the general reaction:

$$R \cdot OH + HO \cdot R \rightarrow R \cdot OR + H_2O$$

This dehydration may be achieved on the large scale by a vapour phase catalyzed reaction, or by reaction with concentrated sulphuric acid.

(d) *Ester formation*

Esters, whose general structural formula is

$$R'CO_2R \quad \text{or} \quad R' \cdot \overset{\textstyle \|}{\underset{\textstyle O}{C}} - OR$$

are formed by reaction of alcohols with *carboxylic acids:*

$$\underset{\substack{\text{Carboxylic}\\\text{acid}}}{R'\!\cdot\!\overset{\displaystyle O}{\overset{\|}{C}}\!\!-\!\!OH} + \underset{\text{Alcohol}}{HO\!\cdot\!R} \rightleftharpoons \underset{\text{Ester}}{R'\!\cdot\!\overset{\displaystyle O}{\overset{\|}{C}}\!\!-\!\!OR} + H_2O$$

The two-part name of an ester indicates first the alcohol fragment, secondly the acid. For example, ethyl acetate is formed by reaction of *ethyl* alcohol with *acetic* acid:

$$\underset{\text{Acetic acid}}{CH_3\!\cdot\!\overset{\displaystyle O}{\overset{\|}{C}}\!\!-\!\!OH} + \underset{\text{Ethyl alcohol}}{HO\!\cdot\!CH_2\!\cdot\!CH_3} \rightleftharpoons$$

$$\underset{\text{Ethyl acetate}}{CH_3\!\cdot\!\overset{\displaystyle O}{\overset{\|}{C}}\!\!-\!\!O\!\cdot\!CH_2\!\cdot\!CH_3} + H_2O$$

These reactions are reversible, ester formation commonly being catalyzed by acid while the reverse *ester hydrolysis* reaction is base catalyzed by, for example, sodium or potassium hydroxide. In ester hydrolysis the reaction results not in the formation of the free acid but of the appropriate salt. Thus the hydrolysis of ethyl acetate in sodium hydroxide is represented by the equation:

$$\underset{\text{Ethyl acetate}}{CH_3\!\cdot\!\overset{\displaystyle O}{\overset{\|}{C}}\!\!-\!\!O\!\cdot\!CH_2\!\cdot\!CH_3} + NaOH \rightarrow$$

$$\underset{\substack{\text{Sodium}\\\text{acetate}}}{CH_3\!\cdot\!\overset{\displaystyle O}{\overset{\|}{C}}\!\!-\!\!ONa} + \underset{\text{Ethyl alcohol}}{HO\!\cdot\!CH_2\!\cdot\!CH_3}$$

(e) *Oxidation*

The products of alcohol oxidation depend on the structure of the alcohol, i.e. on whether it is primary, secondary, or tertiary. Mild oxidations proceed readily in accordance with the following general schemes:

$$R \cdot CH_2OH \xrightarrow{[O]} R \cdot CHO \xrightarrow{[O]} R \cdot CO_2H$$
Primary alcohol Aldehyde Carboxylic acid

$$\underset{\substack{| \\ R'}}{\overset{R}{|}}CH \cdot OH \xrightarrow{[O]} \underset{\substack{| \\ R'}}{\overset{R}{|}}C{=}O \longrightarrow \text{further oxidation unusual}$$
Secondary alcohol Ketone

$$R' - \underset{\substack{| \\ R''}}{\overset{R}{\underset{|}{C}}} \cdot OH \longrightarrow \text{oxidation unusual}$$
Tertiary alcohol

Initial oxidation of both primary and secondary alcohols is formally a dehydrogenation, one molecule of hydrogen being removed. The carbon skeleton of the alcohol is unaltered in the course of these mild oxidations. Aldehydes readily oxidize further, but ketones, like tertiary alcohols, will only oxidize under more severe conditions and in such a manner that the carbon skeleton is broken up into smaller fragments. The industrial oxidation of alcohols is carried out by passing vaporized alcohol and air over silver.

Aldehydes and *ketones* are closely related chemical types, as we might anticipate from their structures:

$$\underset{\substack{| \\ H}}{\overset{R}{|}}C{=}O \text{ or } R \cdot CHO \qquad \underset{\substack{| \\ R'}}{\overset{R}{|}}C{=}O \text{ or } R \cdot CO \cdot R'$$
Aldehyde Ketone

The chemical behaviour of these compounds is dominated by the carbonyl $C{=}O$ group. Carbonyl compounds figure in a number of industrial syntheses; two of the ketone series, acetone, $CH_3 \cdot CO \cdot CH_3$, and methyl ethyl ketone,

$CH_3 \cdot CO \cdot CH_2 \cdot CH_3$, are also important solvents. Alcohol oxidation is not the sole industrial means of obtaining carbonyls in bulk; direct catalytic oxidation of alkenes is now in use, ethylene giving acetaldehyde, propylene giving acetone, and so on. Acetone is also produced by the cumene process (Section 22.5).

The above reactions illustrate the central position which alcohols hold in aliphatic chemistry. It is useful at this point to summarize the relationship which the more important aliphatic compounds have to one another:

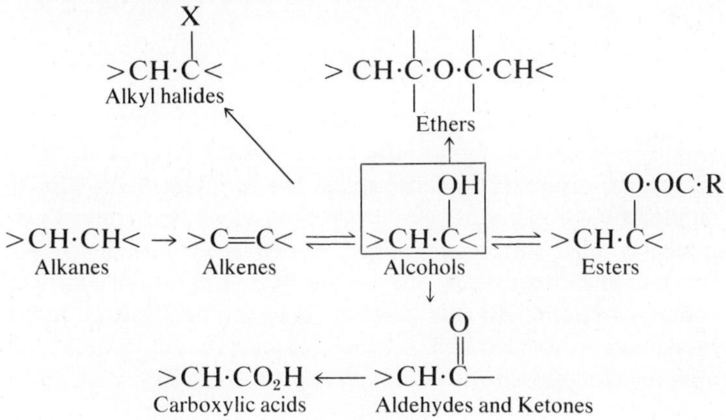

Phenols

22.4. Structure and nomenclature

Phenols have the general formula $Ar \cdot OH$. When a hydroxyl group is attached directly to an aromatic nucleus, the hydrogen atom becomes distinctly acidic. This is perhaps the outstanding difference between phenols and alcohols, although of course the two classes reveal the normal range of differences which exist between aliphatic and aromatic compounds. The presence of the hydroxyl function confers on many phenols a limited degree of water solubility.

Carbolic acid was the name once given to the parent member of the series, phenol, reflecting the nature of the group as a whole. Phenol itself is quite the most important member of the class to which it lends its name. It and a few of its homologues are shown below.

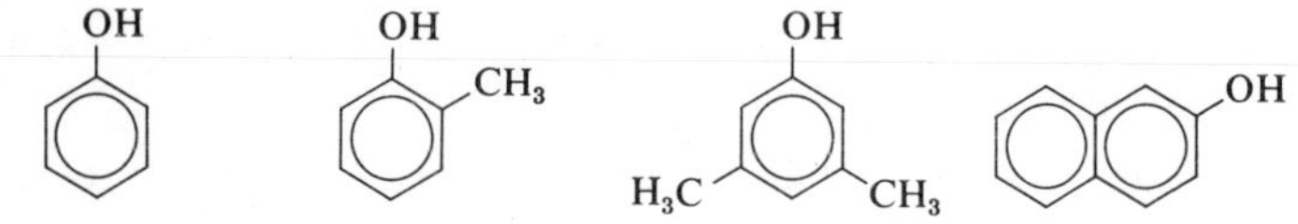

Phenol   *o*-Cresol   3,5-Xylenol   2-Naphthol

22.5. Synthesis of phenols

Coal tar (Section 21.2) includes a phenolic fraction from which is extracted phenol, the cresols, and the xylenols. This traditional source of phenol itself has for long been supplemented by a number of synthetic routes which, in the main, use benzene as a starting material. A petroleum-based process, the cumene process, is one of the more important synthetic routes to phenol. In this process, benzene is alkylated to cumene (isopropyl benzene) which is then readily oxidized to a mixture of phenol and acetone:

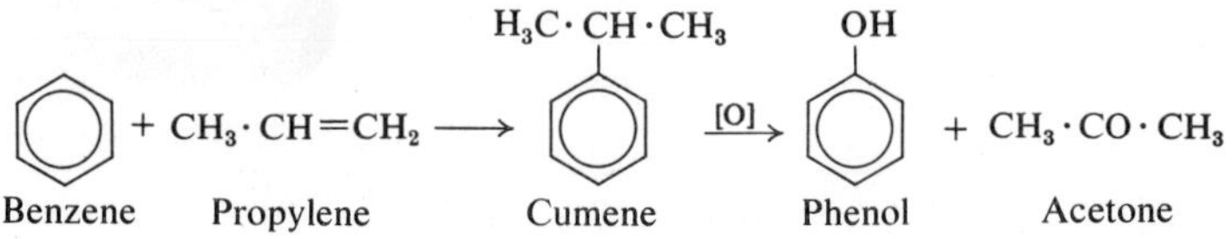

Benzene   Propylene   Cumene   Phenol   Acetone

22.6. Reactions of phenols

Parallels can be drawn between the reactions of alcohols and those of phenols. The latter, for example, are capable of forming *aryl esters* and *aryl ethers* in reactions analogous to those of the alcohols. The comparison need not be carried very far, however, before it becomes clear that we are dealing with two quite distinct classes of compound. Two general reactions illustrate this.

(a) *Acidic behaviour*

Phenols react with strong alkalis in a classical acid-base fashion to form salts and water. Thus phenol reacts with sodium hydroxide according to the following equation:

These salts exhibit typical behaviour, dissociating to ionic species in water to give conducting solutions.

From the above equation we see that any simple phenol is a monobasic acid, and that its dissociation in water can be represented by the equation:

$$Ar \cdot OH + H_2O \rightleftharpoons ArO^- + H_3O^+$$

and thus its dissociation constant K_a is given by:

$$K_a = \frac{[Ar \cdot O^-][H_3O^+]}{[Ar \cdot OH]}$$

The K_a values for a range of phenols clearly indicate that they are very weak acids when compared to the traditional range of inorganic acids; phenol itself has a K_a value of $1 \cdot 2 \times 10^{-10}$ at 25°C.

(b) *Nuclear reactions*

Phenols undergo typical aromatic substitution reactions. The aromatic ring is highly activated by the presence of a phenolic group, and *o,p*-substitution takes place very readily. Paradoxically, the supremely important example here is not provided by any of the better known aromatic substitutions, but by the reaction of phenol with formaldehyde.

Phenol + Formaldehyde → *o,p*-Hydroxybenzyl alcohol + higher substitutions

The reaction is the first step in the production of a wide range of phenol–formaldehyde resins.

CHAPTER 23

Organic Acids and Bases

Carboxylic and Sulphonic Acids

23.1. Carboxylic acids

The compounds represented by the general structural formula

$$R—C \overset{O}{\underset{OH}{\big\langle}} \quad \text{or} \quad Ar—C \overset{O}{\underset{OH}{\big\langle}}$$

are the most important of the organic acids. These are the carboxylic acids. They are capable of donating a proton from the hydroxyl group to suitable acceptors, and are therefore acids in the classical sense.

Like the alcohols, the lower carboxylic acids are freely water soluble because of structural similarities to water, but this solubility decreases with increasing size of hydrocarbon radical. They dissociate in water according to the equation:

$$R \cdot CO_2H + H_2O \rightleftharpoons R \cdot CO_2^- + H_3O^+$$

whence

$$K_a = \frac{[R \cdot CO_2^-][H_3O^+]}{[R \cdot CO_2H]}$$

The K_a values quoted in Table 23.1, which lists some common carboxylic acids, indicate that these are weak acids compared to the inorganic acids, but much stronger acids than phenols. It will be seen that substituent groups can affect the acid strength of both aliphatic and aromatic acids quite markedly. In the examples shown, chloracetic acid and *p*-nitrobenzoic acid each contain a substituent group which exercises a pronounced electron attracting effect, the result of which is to make the carboxylic acid function more positive than in the unsubstituted structure, and therefore more able to split off a proton.

TABLE 23.1. CARBOXYLIC ACIDS

Name	Structure	$K_b(25°C$
Formic acid	$H \cdot CO_2H$	$1 \cdot 77 \times 10^{-4}$
Acetic acid	$CH_3 \cdot CO_2H$	$1 \cdot 75 \times 10^{-5}$
Chloracetic acid	$ClCH_2 \cdot CO_2H$	$1 \cdot 55 \times 10^{-3}$
Propionic acid	$CH_3 \cdot CH_2 \cdot CO_2H$	$1 \cdot 4 \times 10^{-5}$
Butyric acid	$CH_3 \cdot CH_2 \cdot CH_2 \cdot CO_2H$	$1 \cdot 5 \times 10^{-5}$
Benzoic acid	$C_6H_5 \cdot CO_2H$	$6 \cdot 4 \times 10^{-5}$
p-Nitrobenzoic acid	$p\text{-}NO_2 \cdot C_6H_4 \cdot CO_2H$	$4 \cdot 0 \times 10^{-4}$

23.2.   Synthesis of carboxylic acids

The oxidation of aldehydes (Section 22.3) is an easy route to carboxylic acids; acetaldehyde oxidation is the most important industrial example of this method.

$$CH_3 \cdot CHO \xrightarrow{[0]} CH_3 \cdot CO_2H$$

Acetaldehyde   Acetic acid

Vinegar, which is about a 5 per cent aqueous solution of acetic acid, is made by enzyme-induced air oxidation of ethyl alcohol. This reaction can be the ruination of wines and beers, producing as it does a sour acidic flavour in the liquid.

Glyceride hydrolysis provides a suitable route to a group of carboxylic acids which would be difficult to obtain by other methods. Glycerides are naturally occurring esters in which the acid fragments contain long carbon chains. On hydrolysis they yield long-chain acids, the most important of which are listed in Table 23.2.

TABLE 23.2. CARBOXYLIC ACIDS
FROM GLYCERIDES

Name	Formula
Palmitic acid	$C_{15}H_{31} \cdot CO_2H$
Stearic acid	$C_{17}H_{35} \cdot CO_2H$
Oleic acid	$C_{17}H_{33} \cdot CO_2H$
Linoleic acid	$C_{17}H_{31} \cdot CO_2H$
Linolenic acid	$C_{17}H_{29} \cdot CO_2H$

Alkyl side chains of any length on aromatic nuclei are readily oxidized to carboxyl functions, this being the most useful method of obtaining aromatic acids.

Toluene → Benzoic acid

23.3. Reactions of carboxylic acids

The dominating feature of the carboxylic acids is their orthodox acidic behaviour, typified by the formation of ionizable, water-soluble salts on reaction with bases.

e.g. Benzoic acid + NH₃ → Ammonium benzoate

Other reactions important from the industrial standpoint are ester formation (Section 22.3) and reduction to alcohols. This latter reaction is used in industry for the production of long-chain alcohols employed chiefly in synthetic detergent manufacture. The feedstock is a long chain fatty acid mixture obtained from glycerides. The hydrogenation, which is catalyzed, is carried out at high pressures and temperatures.

$$R \cdot CO_2H \xrightarrow{H_2} R \cdot CH_2OH$$

23.4.   Sulphonic acids

Only the aromatic sulphonic acids are of any consequence; these are prepared by direct sulphonation of the aromatic nucleus (Section 21.5). They have the general formula

$$Ar \cdot SO_3H \quad or \quad Ar{-}\overset{\displaystyle O}{\underset{\displaystyle O}{\overset{\|}{\underset{\|}{S}}}}{-}OH$$

and readily donate a proton from the hydroxyl group.

Dissociation of sulphonic acids in water is virtually complete:

$$Ar \cdot SO_3H + H_2O \rightleftharpoons Ar \cdot SO_3{}^- + H_3O^+$$

They are therefore by far the strongest of organic acids, and are more comparable in acid behaviour to sulphuric acid than to the carboxylic acids. Nuclear reactions can occur; the ring is deactivated by the unsaturated sulphonic acid group and *m*-orientation is the principal result of disubstitution.

Sulphonation is much used to render dyestuffs and drugs water soluble, for sulphonic acids dissolve easily in water and other polar solvents. The major outlet of sulphonic acid derivatives is in the synthetic detergent industry (Chapter 28).

Organic Bases

23.5.   Amines

As carboxylic acids are the principal acids of organic chemistry, so amines are the principal bases. Amines are related to ammonia, and their position in the relationship is used as a method of sub-classification:

$$
NH_3 \qquad R{-}NH_2 \qquad R{-}\overset{\displaystyle R'}{\overset{|}{N}}H \qquad R{-}\overset{\displaystyle R'}{\underset{\displaystyle R''}{\overset{|}{\underset{|}{N}}}}
$$

Ammonia   Primary amine   Secondary amine   Tertiary amine

The feature common to all amines is a nitrogen atom which behaves as a basic function for reasons which we shall presently examine. Three typical amine structures are shown below:

$$CH_3 \cdot NH_2 \qquad (CH_3 \cdot CH_2)_2 \cdot NH$$
Methylamine   Diethylamine

NH₂ on benzene ring — Aniline

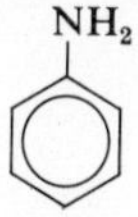

23.6. Synthesis of amines

Aromatic primary amines are obtained by reduction of the corresponding nitro compound,

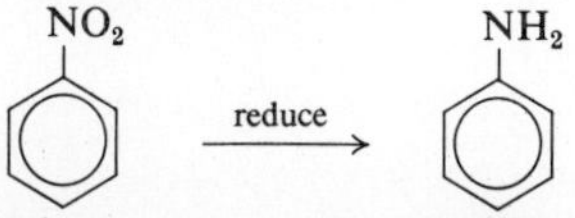

NO₂ on benzene ring (Nitrobenzene) $\xrightarrow{\text{reduce}}$ NH₂ on benzene ring (Aniline)

Secondary and tertiary amines are readily prepared from the primary amine.

Aliphatic amines may be obtained from alcohols, which react in consecutive stages with ammonia to yield eventually tertiary amines. The presence of a large excess of ammonia ensures that the primary amine is the principal product:

e.g. $3CH_3 \cdot CH_2OH + NH_3 \longrightarrow CH_3 \cdot CH_2 \cdot NH_2 \longrightarrow$
    Ethyl alcohol   Ethylamine

$(CH_3 \cdot CH_2)_2 \cdot NH \longrightarrow (CH_3 \cdot CH_2)_3 \cdot N$
    Diethylamine   Triethylamine

Higher aliphatic amines are better prepared from the corresponding alkyl halides by reaction with ammonia.

23.7. Reactions of amines

Amines are basic. This can best be illustrated if we consider the dissociation of a water-soluble amine. This system equilibrates according to the equation:

$$R \cdot NH_2 + H_2O \rightleftharpoons RNH_3^+ + OH^-$$

in which the amine is seen as a classical base, acting as a proton acceptor. The equilibrium constant K_b for the dissociation is given by

$$K_b = \frac{[RNH_3^+][OH^-]}{[R \cdot NH_2]}$$

As with K_a values for organic acids, K_b values vary with amine structure. Some are listed in Table 23.3, from which it will be seen that the aliphatic amines are more basic than ammonia, the aromatic amines less so.

We should expect amines to undergo normal acid-base reactions, and this is the case. Thus water-soluble ionic salts are formed with acids:

cf. $H \cdot NH_2 + HCl \longrightarrow H \cdot NH_3^+Cl^-$
     Ammonia   Ammonium chloride

Salts can also be formed with carboxylic acids; these in turn can be converted to substituted amides. This reaction sequence is made use of in nylon production.

$R \cdot CO_2H + H_2N \cdot R' \longrightarrow R \cdot CO_2^- + H_3^+N \cdot R' \longrightarrow$
  Acid   Base   Salt

$R \cdot CO \cdot HN \cdot R' + H_2O$
Substituted amide

TABLE 23.3. AMINES

Name	Formula	K_b(25°C)
Methylamine	$CH_3 \cdot NH_2$	$4 \cdot 4 \times 10^{-4}$
Dimethylamine	$(CH_3)_2 \cdot NH$	$5 \cdot 1 \times 10^{-4}$
Trimethylamine	$(CH_3)_3 \cdot N$	$0 \cdot 6 \times 10^{-4}$
Ethylamine	$CH_3 \cdot CH_2 \cdot NH_2$	$4 \cdot 7 \times 10^{-4}$
Ammonia	NH_3	$1 \cdot 8 \times 10^{-5}$
Aniline	$C_6H_5 \cdot NH_2$	$4 \cdot 2 \times 10^{-10}$
p-Toluidine	$p\text{-}CH_3 \cdot C_6H_4 \cdot NH_2$	$1 \cdot 2 \times 10^{-9}$

Amines take part in a wide range of reactions over and above those in which they behave as orthodox bases. One example which is of interest within the context of synthetic detergents (Chapter 28) is the sequence of reactions which a primary amine undergoes with an alkyl halide to produce eventually a *quaternary ammonium salt:*

$$R \cdot NH_2 \xrightarrow{RX} R_2 \cdot NH \xrightarrow{RX} R_3 N \xrightarrow{RX} R_4 N^+ X^-$$

Quaternary
ammonium salt

CHAPTER 24

Coal

THE fuel consumption of any society is directly related to its standard of living, and currently we are seeing an increase in world fuel demands so large as to be unimaginable even 40 years ago.

Any system which alters exothermically is a prospective fuel. Nuclear fuels supply heat not from orthodox chemical reactions, but from subatomic changes in the nucleus. These changes are dealt with in Chapter 2, where attention is drawn to the great promise that nuclear energy offers. We still, nevertheless, depend almost entirely on traditional fossil fuel sources, which release their potential energy by exothermic reaction with aerial oxygen.

Fuels must be available in very large quantities at as cheap a price as possible. This has always been a primary requirement, and it always will be. Men first used *wood* as an answer to their fuel needs, and deforestation on a massive scale occurred in many areas over long periods. *Peat* where available was also a convenient fuel. *Coal* became important with the very great increase in fuel demands which accompanied the industrial revolutions. *Petroleum* (including natural gas), a relative newcomer to the world fuel scene, is in process of becoming the dominant member of this group.

In this chapter we shall concern ourselves with coal and its near relations. This fuel, like the other major fossil fuels, is also a valuable chemical raw material, and should always be so regarded.

24.1. Chemical constitution of coal

'Coal' is a term applied to a range of mineral materials which the geologist would regard as organic rock types. Coals, then, vary one from another. Moreover, any given coal cannot be represented by a uniquely defined chemical structure; a single coal is itself a mixture of complex chemical structures.

Raw coal we can conveniently regard as being a three-component material:

$$\text{'Raw' coal} \longrightarrow \begin{cases} \text{'Pure' coal} \\ \text{Inorganic material (mineral matter)} \\ \text{Moisture} \end{cases}$$

We are here using the term 'pure' coal to cover the bulk of the material, which is organic in type and constitutes the essential fuel. The remaining two components have no fuel value as such, and may be regarded as impurities.

The organic material in coal, derived from the decay of woody tissue and other vegetable matter over very long periods of time of up to about 300 million years, contains carbon and hydrogen along with smaller quantities of oxygen, sulphur and nitrogen. The manner in which these elements are ordered in coal structures poses a fascinating problem, and the whole question of coal constitution is still a very open one.

Coal is a polymeric material, i.e. it consists of many high molecular weight molecules, all similar in chemical structure but not necessarily identical. This very definition of polymeric material, whether applied to coal or any other natural or synthetic polymer, implies that we cannot speak of a specific coal structure but only of the structural features which predominate. Coal, containing largely carbon and hydrogen, is chiefly a polymeric hydrocarbon. Much of this hydrocarbon material is aromatic, as indicated by carbon: hydrogen ratios in coal and by the aromatic nature of, for example, coal carbonization and coal oxidation products; but some fully saturated alicyclic rings are also thought to be present. These part aromatic, part alicyclic ring systems exist as condensed clusters (hydroaromatic clusters) which in turn are linked together by short aliphatic bridges, probably methylene $—CH_2—$ bridges mainly.

How do other elements fit into this type of structural unit? Oxygen certainly appears as a phenolic function, as a carbonyl function, and in geologically young coals as a carboxylic acid function. Nitrogen and even more so sulphur are less well defined, but they are likely to occur in heterocyclic structures forming part of the main ring clusters. Partial evidence for this latter view is the appearance of heterocyclic compounds among the products of coal carbonization.

D. W. van Krevelen, an outstanding contributor to the field of coal constitution, has emphasized that coal structure cannot be tied to a visual model but should be expressed as a series of structural parameters. Nevertheless, a visible model helps all of us to get a clearer picture of what a segment of a coal molecule might look like. One such model is reproduced in Fig. 24.1. It attempts to represent a segment of a coal of

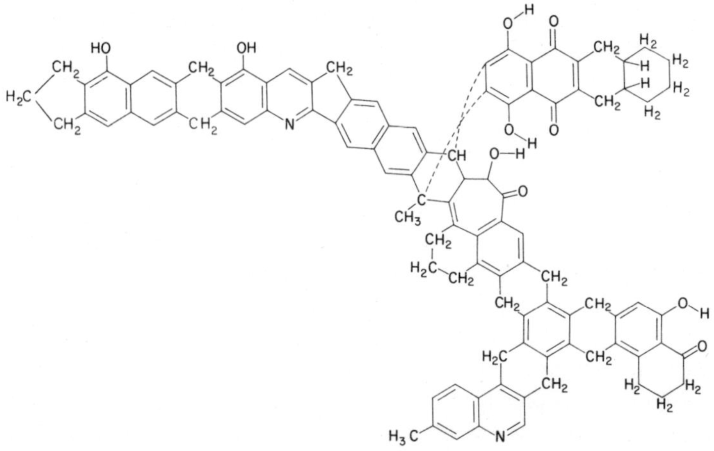

FIG. 24.1. Representation of a coal 'molecule'. (Given, P. H., *Nature*, **184**, 980 (1959).)

specific analysis, and does not therefore include every structural feature mentioned above; thus no sulphur appears, nor any carboxylic acid function.

The aromatic nature of coal helps to explain why it is a solid. The ring clusters tend to be planar, and packing of these clusters or lamellae is efficient enough to produce a solid mass with a high softening temperature. This contrasts with petroleum, where the large number of flexible aliphatic structures leads to a low packing efficiency and a liquid phase.

24.2. Analysis of coal

The complex nature of coal rules out, for many purposes, analytical techniques which measure precisely some clearly defined chemical or physico-chemical property. Some of the more important tests are empirical; they do not measure absolute functions, and if meaningful results are to be obtained such empirical tests must be carried out according to standard specified conditions. In the U.K. the British Standards Institution is responsible for the standardization of coal and coke testing, the various procedures being embodied in B.S. 1016 and 1017. The equivalent body in the U.S.A. is the American Society for Testing Materials (A.S.T.M.).

(a) *Proximate analysis*

Perhaps the most important coal test, the proximate analysis, records three parameters: *moisture, volatile matter* and *ash*, the residue being reported as *fixed carbon*.

Coal is a hygroscopic material and moisture is always present to a small extent. This hygroscopic moisture is usually referred to as *inherent moisture*, being the equilibrium moisture content of a coal sample air-dried in the laboratory prior to proximate analysis. Coal may also contain varying quantities of *free moisture*, which can readily be determined, but must not be confused with the proximate analysis figure. Moisture in coal contributes nothing to the fuel potential of the material; but damp coal burns more evenly on a fuel bed than dry coal, and for this reason a coal feed to a furnace may be sprayed with water to a predetermined moisture content before firing.

Volatile matter is an important coal parameter, and goes a long way towards classifying a given coal. It is defined as the percentage weight loss from a coal sample heated for 7 minutes at $900°C \pm 5°C$ (precise conditions vary a little in different countries).

Ash in coal is determined by burning off the organic material and weighing the residue. This ash residue cannot be directly equated with original mineral matter of the coal. During the process of combustion, carbonates will decompose to oxides, which together with other like transformations will yield an ash figure perhaps quite different from the true mineral matter figure. The latter can be determined indirectly, but laboriously, and if the ash figure is less than 5 per cent this conversion is usually avoided. Ash in coal is, like moisture, of no fuel value but it does have the virtue of protecting grate bars from direct contact with incandescent carbon, a contact which would certainly lead to totally unacceptable corrosion rates. Ash can be troublesome if it has too low an *ash fusion temperature*, for ready softening at high temperatures leads to clinker formation on cooling. Clinker on grates interferes with combustion air supply, is difficult to discharge, and is very undesirable.

If proximate analyses for different coals are to be compared, this is best done by reducing the analytical figures to a dry, ash-free (d.a.f.) basis, i.e. by eliminating the "impurities" from the analyses, and simply stating volatile matter against fixed carbon. Otherwise the presence of different levels of ash and moisture in two coals may obscure very real differences or similarities which exist between the materials. In very accurate work, a coal analysis may ultimately be quoted on a d.mm.f. (dry, mineral-matter free) basis.

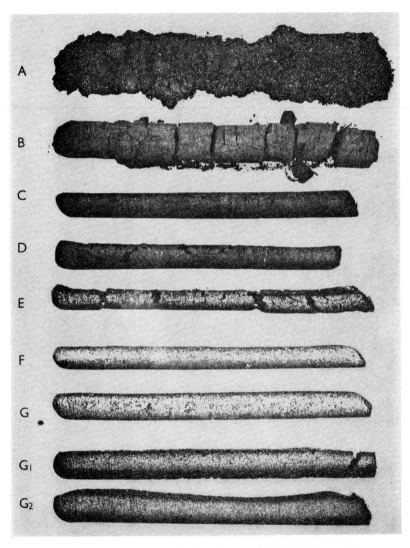

FIG. 24.2. Gray–King standard coke types. (B.S. 1016, part 12, 1959. Reproduced by permission of the British Standard Institution, London, W.1. The figures are not to scale, and are unsuitable for laboratory use).

(b) *Ultimate analysis*

The ultimate analysis of coal is not an empirical test, but a precise determination of the elemental constituents using standard analytical techniques. The test, as defined by the British Standard, determines carbon, hydrogen, nitrogen and sulphur; oxygen may be determined by difference. Although not considered part of the ultimate analysis, methods are also specified for determining trace elements, in particular chlorine, phosphorus and arsenic.

(c) *Gray–King assay*

Each year large quantities of coal are carbonized in process of making metallurgical coke, coal gas, and other products. In the U.K. the behaviour of a coal during carbonization is assessed by means of a Gray–King assay. The standard low temperature assay is carried out by carbonizing a sample of coal at a maximum temperature of 600°C, the various decomposition products being collected as required.

The real importance of the Gray–King assay is its use as a British Standard test for classifying cokes. By comparison with a series of photographs of standard cokes, shown in Fig. 24.2, the coke residue from the assay is accorded a classification which runs from A to G, and thence from G_1 to G_9 type cokes. Thus an effort is made to indicate the two chief properties of a coke, viz. the degree to which it has swollen, and the degree to which individual particles have agglutinated to form one mass. The significance of key points on the scale are indicated below:

Coke type:	A	G	G_9
		hard	
Description:	pulverent	no swelling or shrinking	highly
	highly swollen	(standard coke)	swollen

(d) *Calorific value*

The calorific value, or heat content, of any fuel is of self-evident importance. For solid and liquid fuels, it is determined by burning a sample in a steel calorimeter under 25 atm of oxygen, and measuring the heat output in a surrounding calorimeter. For coals, calorific values range from about 9000 to 16,000 Btu per pound (5000–9000 cal per gram).

24.3. Classification of coal

We need only remind ourselves of the highly complex chemical structures which are believed to occur in coals to realize that rigorous classification of these materials is likely to be very difficult.

(a) *Coal rank: the Seyler classification*

The *rank* of a coal denotes its geológical age, i.e. its position on the scale describing the change of vegetable materials through various fairly well-defined stages to the final product of the coalification process, graphite:

```
Vegetable ⟶ Peat ⟶
material
Low rank                      true coals
                                 |
                    ┌─────────────────────────────┐
                    Lignites ⟶ Bituminous ⟶ Anthracites ⟶ Graphite
                                 coals                      High rank
```

Change in rank is reflected by a steady change in other parameters, notably carbon content and volatile matter, but also hydrogen content, oxygen content, calorific value, and inherent moisture content. The progression outlined above is not in itself comprehensive enough to serve as a classification system, but most of the more detailed schemes in use are based on rank changes, and are commonly formed by plotting or tabulating two or more parameters which are themselves functions of rank, and so obtaining a correlation over a wide range of coals.

We can see this clearly in the most detailed of the classification schemes devised for British coals, that due to Seyler. Here carbon and hydrogen contents for a series of coals are plotted to obtain a *coal band*, which is subdivided into a number of groups. The primary divisions in the Seyler scheme are shown in Table 24.1.

TABLE 24.1. DIVISIONS IN THE SEYLER
CLASSIFICATION SCHEME

Coal type	%C	%H
Peat (average)	56·5	5·5
Lignite	65–76	5·25–5·1
Ortho-lignitious	76–80	5·1 –5·05
Meta-lignitious	80–84	5·05–5·0
Para-bituminous	84–87	5·0 –4·95
Ortho-bituminous	87–89	4·95–4·8
Meta-bituminous	89–91	4·8 –4·25
Carbonaceous	91–93	4·25–3·3
Anthracite	93–94·8	3·3 –2·5

Having constructed the basic coal band, other relationships can be represented within the Seyler scheme (as indeed in other classification schemes). Thus lines of constant volatile matter and constant calorific value are readily superimposed on the initial chart. Much information beyond this is given in the fullest form of the Seyler chart, which is presented and discussed in some detail elsewhere.† A restricted version is reproduced in Fig. 24.3.

Coals in the main are derived from woody tissues, but small quantities of fossilized material from other vegetable sources also occur. If too much of this latter material is present, an abnormal coal which does not fall within the coal band of the Seyler chart may result. For such abnormal or *dull* coals corrections have to be made in applying the Seyler chart to them. A majority of British coals are normal, or *bright*, coals.

†Ed. Spiers, H.M., *Technical Data of Fuel*, World Power Conference, London, 1961.

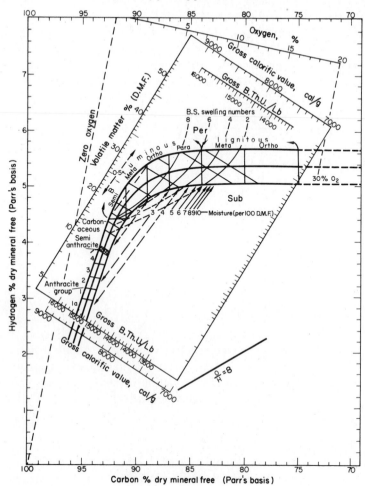

Fig. 24.3. The Seyler chart.

(b) National Coal Board classification

The complexity of the Seyler classification makes it unwieldy for many practical purposes. The National Coal Board now uses a scheme in which two readily determined and very practical parameters, viz. volatile matter and Gray–King coke type, form the basis of a division of British coals into nine primary groups, numbered 100 to 900. (These must not be confused with the division of coals into five groups for the domestic and commercial market.) The groups are in turn subdivided into classes, 101 and 102, 201, and so on, which may be split yet again into, e.g. sub-classes 201*a* and 201*b*. Separation of different coals within the National Coal Board scheme is not always satisfactory, but it does suffice for most purposes. The scheme is summarized in Table 24.2(a) and 24.2(b).

Several other classification schemes have been proposed and are similar to those already described in as far as they relate general properties within the framework of a two- or three-parameter plot. The discussion of this subject set out by Francis† includes coverage of American and European systems.

24.4. Coal combustion

Coal may be burned in industrial furnaces on a number of different types of grate. The reactions which occur are essentially independent of the type of grate in use; indeed many of the phenomena occurring in a large industrial installation differ from those in a domestic grate only in magnitude.

When coal is fed to an incandescent fuel bed it decomposes. Combustible volatile matter is evolved, and a coke residue is left on the bed itself. We must therefore consider the combustion process to be two-stage, the first stage occurring by interaction of primary air with coke in the fuel bed, the second occurring by interaction of secondary air with volatile combustibles in the vapour space above the bed. We shall deal with the more important features of primary and secondary combustion in turn.

†Francis, W., *Coal: its Formation and Composition*, Arnold, London, 1961.

TABLE 24.2. NATIONAL COAL BOARD CLASSIFICATION OF BRITISH COALS

(a) Anthracites, Low-volatile Steam Coals and Medium-volatile Coals

Group, $VM_{d.mm.f.}$ and General description	Class	Sub-class	$VM_{d.mm.f.}$	Gray–King coke type	Specific description
100 Under 9·1 Anthracites	101		Under 6·1	A	Anthracites
	102		6·1–9·0	A	
200 9·1–19·5	201		9·1–13·5	A–C	Dry steam coals
		201a	9·1–11·5	A–B	
		201b	11·6–13·5	B–C	
Low-volatile steam coals	202		13·6–15·0	B–G	Coking steam coals
	203		15·1–17·0	E–G$_4$	
	204		17·1–19·5	G$_1$–G$_8$	
	206		9·1–19·5	A–B (VM up to 15·0) A–D (VM above 15·0)	Heat-altered low-volatile steam coals
300 19·6–32·0	301		19·6–32·0	G$_4$ and over	Prime coking coals
		301a	19·6–27·5		
		301b	27·6–32·0		
Medium-volatile coals	305		19·6–32·0	G–G$_3$	(Mainly) heat-altered medium-volatile coals
	306		19·6–32·0	A–F	

(b) *High-volatile Coals* ($VM_{d.mm.f.}$ above 32·0%)

Group	$VM_{d.mm.f.}$		Gray-King coke type	Description
	32·1–36·0	Over 36·0		
	Class			
400	401	402	G_9 and over	Very strongly caking
500	501	502	G_5–G_8	Strongly caking
600	601	602	G_1–G_4	Medium caking
700	701	702	E–G	Weakly caking
800	801	802	C–D	Very weakly caking
900	901	902	A–B	Non-caking

Primary combustion is best illustrated by considering a typical system in which fresh fuel is overfed to the top of the bed. Decomposition occurs rapidly, and the coke residue then moves down through the bed, reacting as it travels with primary air rising up through the grate bars. This counter-current system is a heterogeneous one, being the reaction of a gas at a solid surface. In the ideal case reaction will be complete before solid combustible material reaches the grate bars, at which point the residue should consist solely of inorganic ash. In practice this is very difficult to achieve, and a little unburned carbon is usually rejected with the ash.

If gas samples are taken through a fuel bed of the type described and subsequently analyzed, a profile is obtained in which distinct zones emerge. This is shown in Fig. 24.4, where changes in gas content against bed depth are plotted, and Fig. 24.5, which interprets these gas changes diagrammatically.

Air rising through the inert *ash zone* first reacts with carbon in an *oxidation zone* to produce mainly carbon dioxide. (Formation of the dioxide is known to be a two-step process, the first step being a heterogeneous reaction of oxygen at a carbon surface to yield carbon monoxide. This product subsequently

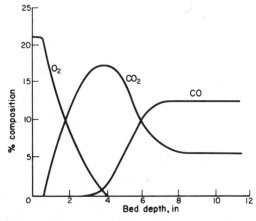

Fig. 24.4. Variation of gas analysis with combustion bed depth.

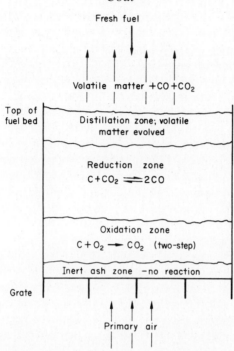

Fresh fuel

Volatile matter $+CO+CO_2$

Top of fuel bed

Distillation zone; volatile matter evolved

Reduction zone
$C+CO_2 \rightleftharpoons 2CO$

Oxidation zone
$C + O_2 \longrightarrow CO_2$ (two-step)

Inert ash zone —no reaction

Grate

Primary air

FIG. 24.5. Reaction zones in a combustion bed.

diffuses from the carbon surface and reacts very rapidly with further oxygen to yield carbon dioxide.) Reaction is virtually complete in the oxidation zone, all oxygen being consumed. Carbon dioxide then rises through the upper reaches of the bed, and in so doing reacts with further quantities of carbon to form carbon monoxide — hence the defining of a *reduction zone*. This latter zone is essentially the region in which the important gas – solid system carbon–carbon monoxide–carbon dioxide attempts to reach equilibrium. The degree to which equilibrium is approached is determined chiefly by the residence time in the bed; carbon monoxide is generally the principal product. The top of the fuel bed is called the *distillation zone* wherein occurs the evolution of volatile matter from fresh coal decomposition.

We can now see that from the upper surface of the fuel bed there emerges not only volatile matter, but also large quantities of carbon monoxide and some carbon dioxide, products of the primary combustion process. The escape of unburned carbon monoxide from a furnace cannot be tolerated because of the high potential heat loss which would result. Thus the process of secondary combustion, in which the gaseous products of the primary procedures occurring in the fuel bed should be completely oxidized, is in a sense the critical stage in determining the overall combustion efficiency of a furnace.

Secondary combustion is a homogeneous gas-phase reaction in which the end-products should be carbon dioxide and water vapour. If the burning gases are ejected too quickly from the furnace into cooler regions, combustion may be interrupted. Injection of preheated secondary air into the combustion chamber is therefore so designed that a high degree of turbulence and consequently rapid mixing of the gases results. If the gases are accidently cooled below their ignition temperature, elemental carbon will appear as black sooty smoke at the chimney head. It is difficult to avoid soot emission when starting a furnace from cold, hence in the U.K. the Clean Air Act permits industrial furnaces to make smoke for 15 minutes after preliminary firing. Properly operated, a hot furnace which nowadays will most likely be automatically stoked should not encounter soot-emission troubles.

Smoke emission of a different sort can occur if the secondary air supply is insufficient. In this case unburned gases will escape, and this will be signalled by brown smoke from volatile hydrocarbon material appearing at the chimney head. In such instances, however, the big loss in potential heat is due not to the visible hydrocarbon loss, but to its invisible companion, carbon monoxide.

Air supplies to a furnace have to be carefully controlled if maximum efficiency is to be attained. Both primary and secondary air streams must be adequate enough to ensure complete combustion at their respective sites. Complete combustion can never be achieved with theoretical quantities of air, and an excess must always be present. We must take account of another heat loss, however, when dealing with combustion air supply. Waste gases will be ejected at a chimney head at a temperature higher than ambient temperature, and this effluent stream therefore constitutes a considerable sensible heat loss. Most of this heat loss is due to nitrogen, accounting as it does for 79 per cent of the air supply. Any increase in excess air therefore increases substantially the volume of waste gases, and with it the associated sensible heat loss.

A deficiency in air supply means a loss in potential heat due to incomplete combustion. An unnecessary amount of excess air means a loss in sensible heat in the waste gases. These opposing factors must be optimized in any furnace system, as indicated in Fig. 24.6. The optimum excess air figure for a coal-fired furnace may be of the order of 50 per cent. Liquid and gaseous fuels lend themselves to a much higher degree of automated control, and with these fuels greatly reduced excess air figures can be obtained.

Much of the sensible heat in a waste gas stream is recovered by the use of heat exchangers, in which combustion air streams are preheated. In suitable cases fuels may also be preheated. We might reasonably postulate that the more heat that can be extracted in a heat exchanger, i.e. the lower the ejection temperature of the waste gases, the more efficient will be the overall process. Here again, however, a compromise has to be made. Sulphur in the course of fuel combustion oxidizes to a mixture of sulphur dioxide and sulphur trioxide. If a waste gas stream is cooled below its dew point, such that water condensation occurs, the sulphur gases partially dissolve to form highly corrosive acidic solutions. It is essential, therefore, to discharge the effluent gas at a temperature higher than its dew point in order to avoid acid attack on the plant. (The discharge of sulphur gases to atmosphere is itself by no means desirable. Such discharges are controlled by statutory limitation of the sulphur content of the gases.)

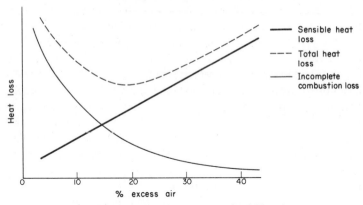

FIG. 24.6. Relations between excess air and heat losses
in a furnace.

24.5. Pulverized fuel

If a solid fuel is ground to a very finely divided state it can be transported in pressurized air streams and generally handled as if it were a fluid. Fluidized particle systems are now important in a number of fields, nowhere more so than in coal-fired furnaces.

Firing a furnace with pulverized coal offers several important advantages. Like any truly fluid system, automation and very fine control become possible. Much less excess air is required compared to a furnace firing on lump coal. Poor-quality coals which would otherwise be unsuitable can be burned. High combustion intensities and high efficiencies can be achieved. These and other advantages have led to a remarkable increase in pulverized fuel firing. This increase has been most noticeable in large installations, particularly coal-fired power stations, because in the past the high cost of grinding a material to sufficiently small size could only be absorbed by very big units. With improvements in the economics of grinding operations, this is now less true.

Pulverized fuel furnaces differ in some ways from orthodox

coal-fired furnaces. There is no grate, the fuel particles (perhaps up to 40 million per cubic inch) being carried on an air stream into a stationary three-dimensional flame through which they travel as they combust. Secondary air is injected into the flame zone at a carefully selected point. The mechanism of combustion, however, parallels that of lump coal in that two distinct stages have been observed. Each particle first devolatilizes, the volatile matter burning round about the minute coke residue. A heterogeneous reaction then occurs between coke particle and oxygen. Thus the first stage is analogous to secondary combustion in an orthodox furnace, the second stage to the processes occurring in primary combustion in a fuel bed.

The high temperatures which are reached in these systems necessitate furnace walls which are water cooled by imbedded tubes. If the ash fusion temperature is low enough, molten ash may result, and this can be tapped off (a 'wet-bottom' furnace). Otherwise ash will be ejected as fly ash in the hot waste gases, and will be separated from them in ancillary units.

24.6. Coal carbonization

Coal decomposes at high temperatures to yield complex gaseous, liquid and solid products in the distillate, and a coke residue. Carbonization of coal may be carried out at temperature maxima of between 700°C and 1350°C, high-temperature processes being the more common.

Coal gas is the name given to the truly gaseous fraction from which has been removed a variety of undesirable impurities, principally hydrogen sulphide, hydrogen cyanide, ammonia, and benzole vapour. A coal gas from high temperature carbonization contains about 50 per cent hydrogen, 25 per cent methane and 10 per cent carbon monoxide, and will have a calorific value of approximately 500 Btu per cubic foot. This gas has been the foundation of towns' gas supplies in the past in areas rich in coal. Coal carbonization is no longer economic, however, as a route to towns' gas when compared with oil-based processes, or with readily available natural gas. In the U.K. we have been witnessing a large and rapid reorganization of the gas industry which has relegated traditional carbonization to a very small role in this sphere.

Coal tar contains in essence the liquid and volatile solid products from degradation of the main coal substance. It has already been described as a rich source of aromatic compounds (Section 21.2). A further outlet provides a range of *coal tar fuels*, these being coal tar fractions, classified according to viscosity, which may be used as liquid fuels in furnace combustion in analogous manner to petroleum fuel oils. Coal tar fuels have lower sulphur contents than corresponding petroleum products, an attractive feature in metallurgical operations where sulphur can have a detrimental effect on the end product.

Coke is, in any carbonization process, the principal product on a weight basis and consists largely of carbon along with the ash residue of the coal and a little moisture. Its properties vary markedly with the type of coal carbonized. *Metallurgical coke*, which in the blast-furnace process for pig-iron production acts both as fuel and reducing agent, must be hard and dense enough to carry a heavy burden without breaking into fines, and must also be low in sulphur and phosphorus content. Good coking coals are now in short supply, and one technique adopted by the iron and steel industries to reduce coke usage has been the introduction of oil injection into blast furnaces. *Gas coke* from the towns' gas industries is softer and more reactive than metallurgical coke, usually having been derived from a high volatile matter bituminous coal. This embarrassing by-product has been used by the gas industry and others to manufacture water gas and producer gas; some gas cokes are also suitable for use as domestic solid smokeless fuels. These latter materials are technically defined as *semi-cokes*, i.e. carbonization residues which have been stripped of smoke-producing hydrocarbons, but which still retain sufficient volatile matter (mainly hydrogen) to maintain combustion in an open grate. The most reactive semi-cokes are made by low-temperature carbonization processes and are marketed under a variety of different brand names (e.g. Coalite and Rexco in the U.K.).

24.7. Coal: a chemical raw material

The importance of coal in the future may lie not in its potential as a solid fuel but as a precursor of other more highly refined chemical materials. We can regard secondary fuels derived from coal as belonging to this class, for such fuels are materials which have been processed to some extent. The distinguishing feature of the fuel-processing industries is that the required end product is not the chemical itself, but its heat or power content.

Coal carbonization was the earliest technique developed to convert the raw material into other products. Consider the wide range of secondary fuels which may be derived by the thermal decomposition of coal: coke, semi-coke, motor benzole, coal tar fuels, coal gas. Producer gas and water gas (Section 11.5) are coal-based products. Carbonization we have noted, moreover, as a route to coal chemicals: coal tar products are obvious, other products perhaps less so; for example, we would hardly think of Neoprene rubber as being a coal chemical, until we realize that Neoprene is made from chloroprene, which is made from acetylene, much of which has been made from calcium carbide, which is made from coke, which is made from coal!

For the production of liquid and gaseous fuels and of chemicals, however, carbonization is falling out of favour. It is not a very flexible process, and therefore not a very efficient one. Newer, more sophisticated processes for the conversion of coal into a range of useful products are available, the more important of which we shall note.

Hydrogenation of coal degrades the main structure and yields a series of liquid products which can replace petroleum fuels. The *Bergius process*, in which a mixture of coal and coal tar is hydrogenated at high pressure and temperature in the presence of a catalyst, was the forerunner of a number of hydrogenation processes which were used during the 1939–45 war to produce fuel oils, diesel oils and gasolines. Coal hydrogenation oils are not at present competitive with petroleum oils.

Synthesis gas manufacture from coal leads to a number of further uses. Synthesis gas, i.e. a mixture of carbon monoxide and hydrogen, can be made as an equimolar mixture by the water gas reaction:

$$C + H_2O \rightleftharpoons CO + H_2$$
Synthesis gas

These gases may subsequently be converted into hydrocarbon mixtures and a wide range of oxygenated compounds (see, for example, Section 22.2) by catalytic processes the best known of which is the *Fischer–Tropsch process*. This latter process can be made astonishingly flexible by altering the reaction variables, especially the molar ratio of carbon monoxide to hydrogen and the catalyst.

The Lurgi process is also used to produce synthesis gas. Here coal is completely gasified by blowing it with a mixture of steam and oxygen at pressures of the order of 25 atm. In the U.K. Lurgi gas is enriched with hydrocarbon gases and distributed as towns' gas. In South Africa, a Lurgi plant is in operation to make synthesis gas for ultimate conversion to liquid fuels. The Lurgi process has become uneconomic either for towns' gas manufacture or liquid fuels synthesis compared to petroleum-based processes, unless special factors such as cheap labour or cheap coal weigh in its favour.

24.8. Future prospects

At almost every point coal has lost ground to petroleum since 1950. It must not be dismissed, however, as a dying concern. Coal remains a cheap fuel in terms of cost per Btu. It suffers technical disadvantages in lacking the fluidity of liquid or gaseous systems, and in presenting a considerable ash disposal problem. The former disadvantage can be overcome by using coal in pulverized form, as already described, a technique which can be applied not only to combustion, but also to gasification process. Fluidized bed gasification of pulverized coal, with all its chemical as well as fuel implications, will possibly become important in the future. Coal can also be gasified underground, *in situ*, by treatment with mixtures of air and steam or oxygen and steam. Underground gasification eliminates mining costs and is very attractive, but poses technical problems which have yet to be solved.

In the U.K. the main outlet for coal in the future is likely to be metallurgical coke production, although generating stations will continue to use it for some time yet. Production, which prior to the North Sea gas finds was about 190 million tons per annum, will fall to a level substantially below this figure by the mid-1970's. The real importance of coal in the future may be as a chemical raw material, an outlet which perhaps has been too readily regarded in the past as of minor importance. A future generation may look back on the era of coal combustion as one of appalling waste.

CHAPTER 25

Petroleum-Based Industries

THERE is a certain inevitability about the place of petroleum in chemical production today which creates the impression that crude petroleum is the sole basis of current organic industrial practice. We must remember that coal or fermented vegetable material can provide alternative raw material sources; it is simple economics which currently favours petroleum. We must also remember that liquid and gaseous petroleum fuels greatly exceed petrochemicals in production volume; petrochemicals proper represent a minor outlet, although a very profitable one, for crude petroleum.

25.1. Chemical nature and classification of petroleum

Petroleum, like coal, is a generic name. Any one oil is a highly complex mixture of (mainly) hydrocarbons, and one oil may be very different from another in its detailed composition. Unlike coal, however, individual compounds can be distinguished in crude oil. Among these we have already noted the very simplest of hydrocarbons, low molecular weight materials like methane and its immediate homologues. Although relatively high molecular weight material does occur, petroleum is not polymeric in the same sense as coal.

Crude petroleum, which has probably arisen from the decomposition of marine organisms, consists of carbon and hydrogen and minor quantities of oxygen, nitrogen and sulphur. The hydrocarbon groups which constitute the bulk of any petroleum oil are *alkanes*, *cycloalkanes* (*naphthenes*) and *aro-*

matics, the latter usually being of minor importance. Alkenes may also occur in very small amounts.

The minor elements in petroleum are difficult to characterize. Oxygen occurs as a carboxylic acid function, particularly as a series of *naphthenic acids* in which the hydrocarbon radical is a naphthene ring. The most important of the always undesirable sulphur compounds are the *mercaptans* or *thioalcohols*, R·SH. Sulphur also occurs in *thioether* —C—S—C— and disulphide —C—S—S—C— linkages, and as *heterocyclic sulphur*. Nitrogen appears in a number of *heterocyclic nitrogen base* structures.

Petroleum crudes do not display the same type of family relationship as coals. The ultimate analyses of a wide range of oils are distributed around mean values of 84·5 per cent carbon, 12·5 per cent hydrogen, and residue minor elements, and variations from these mean values are only of the order of a few per cent. Primary classification is simply based on the nature of the residue obtained on distillation of an oil. Three classes exist:

  (i) *paraffin base oils*, in which the residue is largely paraffin wax;
 (ii) *asphalt base oils*, in which the residue is largely asphalt (or bitumen), i.e. an aromatic residue;
(iii) *mixed base oils*, in which the residue is a mixture of the above two classes.

While the primary nature of an oil determines the manner in which it is refined, individual fractions are in essence defined and classified by physical properties such as boiling range, viscosity and so on. We are here saying, as with coal, that the history of a given fraction or its detailed chemical composition are not of first consequence. If the fraction performs its given task satisfactorily, that is sufficient. We shall soon see, however, that satisfactory performance of, for example, a gasoline can better be achieved by taking account of chemical structure as well as gross physical properties.

25.2. Analysis of petroleum oils

The methods of analyzing petroleum oils are too numerous to consider in detail, but we shall note some of the more important tests. The authoritative manuals in this field are, in the U.K., the Institute of Petroleum handbook *Standard Methods for Testing Petroleum and its Products*; and in the U.S.A., the A.S.T.M. handbook *Standards on Petroleum Products and Lubricants*. Inspection of standard methods reveals that while some of the tests measure specific physical parameters, many of them are empirical. Hence, as with solid fuels, it is important to carry out a determination according to the prescribed method.

(a) *Specific gravity*

This value is determined orthodoxly. It gives some indication of the oil composition, being inversely proportional to the alkane (or paraffin) content of an oil. In the U.S.A. a reciprocal gravity figure, the A.P.I. (American Petroleum Institute) gravity, is often quoted.

(b) *Viscosity*

Oil viscosity is very important in a number of fields. *Kinematic viscosities* are measured in a U-tube viscometer. Often used in the U.K. is the *Redwood viscosity*, which is simply the time of flow of an oil sample through the standard orifice of a Redwood viscometer. The equivalent American unit is the *Saybolt viscosity*. For lubricating oils, the variation of viscosity within a normal range of working temperatures must also be determined, and is expressed by means of a *viscosity index* (Section 28.2).

(c) *Vapour pressure*

Vapour pressure is particularly relevant to motor fuels, indicating as it does the ease of cold-weather ignition, and the danger of vapour lock occurring. The value is determined by the *Reid method*.

(d) *Distillation range*

Again within the field of motor fuels, the distillation range is a primary test. Specific values (e.g. the initial boiling point, the temperatures of 10, 50, and 90 per cent distillate, and the final boiling point) are laid down for gasolines, diesel oils and related fuels.

(e) *Aniline point*

This very simple test in which the solubility in aniline of diesel oils is measured indicates empirically the fraction of alkanes present in the sample. The aniline point is often combined with the A.P.I. gravity and expressed as a *diesel index* (Section 25.12).

(f) *Flash point*

The flash point of an oil (not to be confused with the spontaneous ignition temperature) indicates the temperature of ignition from an open flame. Flash points are important in assessing the fire hazard associated with the bulk storage of liquid fuels.

(g) *Calorific value*

As with solid fuels, a bomb calorimeter is used to measure calorific values of liquid fuels.

(h) *Carbon residue*

The carbon residue of an oil is the coke residue left after all volatiles have been driven off in non-oxidizing conditions.

Carbon deposits can be very troublesome in a number of fuel injection systems, in oil burners, and in lubricating oils; on the other hand carbon residues greater than about 6 per cent are desirable in some oil-based gas-making processes. Carbon residues are measured by either the *Conradson* or the *Ramsbottom* test.

25.3.   Petroleum refining

Petroleum crudes are never used as such. It is the task of a petroleum refinery to divide its raw material into fractions, each tailored to a particular job. A number of refining operations are commonly used, the most important of which is physical separation by distillation. Distillation procedures are fully discussed in Chapter 16.

Initial distillation in the oil industry is carried out as a continuous process operating at atmospheric pressure. Secondary distillation of the heavier ends may be required, depending on refinery demand and the nature of the oil. Where needed, fractionation of high boiling material takes place at reduced pressures in order to lower working temperatures and avoid thermal cracking.

A schematic representation of crude petroleum fractionation is given in Fig. 16.4, which simplifies the whole process and indicates an ease of separation difficult to obtain in practice.

Simple distillation is seldom a self-sufficient process, and a variety of other refining techniques (dealt with in Chapters 16 and 17) are used in developing petroleum end products. These include steam and azeotropic distillation, absorption, solvent extraction, and other methods of separation which are applied in a variety of ways.

In addition to physical methods of separation and refining, the oil industry also employs processes designed to change the chemistry of the products. Into this category would fall cracking procedures of various kinds. Chemical processes are particularly important in petrochemicals production and in gasoline technology.

Of the multiplicity of fractions obtained from crude petroleum refining, we shall consider in detail in the following sections only the more important liquid products, and in particular those which are liable to be of interest to the engineer.

25.4. Gasoline: production

Gasoline, or petrol, is the fuel used in the spark ignition engine. In mild climates a fraction distilling in the range 30–150°C is required. Primary distillation of a crude petroleum will never yield more than about 30 per cent gasoline ('straight-run gasoline'), and sometimes considerably less than this. Market demands are now so severe that straight-run gasoline supplies have to be heavily supplemented by cracking higher boiling fractions.

Thermal cracking of suitable fractions was first used to increase gasoline yields, but cracking in the presence of a catalyst is now favoured. The milder conditions which can be used in *catalytic cracking* give higher yields of a better-quality gasoline and less elemental carbon.

In catalytic cracking, the elemental carbon produced is deposited at the surface of the catalyst, duly rendering it inactive. A regeneration cycle, during which this carbon is burned off, is therefore necessary. Interrupting a production line is expensive, and continuous processes have been developed in catalytic cracking whereby catalyst regeneration can be carried out constantly without interfering with gasoline production. The best way of doing this is to use a *fluidized bed* system, a diagram of which is shown in Fig. 25.1. The key feature of this type of operation is that finely divided solid material, if kept agitated by a liquid or gaseous stream, can be handled as part of a fluid system, i.e. it can be pumped and transported as flexibly as a true fluid. The catalyst bed is 'fluidized' by the upward passage of oil feed vapour. Spent catalyst is drawn continuously from the bottom of the cracking chamber, transported on an air stream to the regeneration chamber in which the carbon

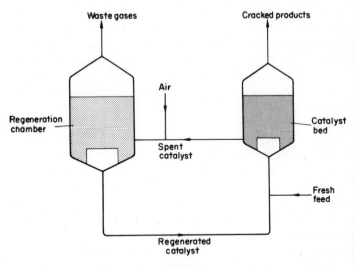

FIG. 25.1. Schematic representation of a fluidized bed cracking plant.

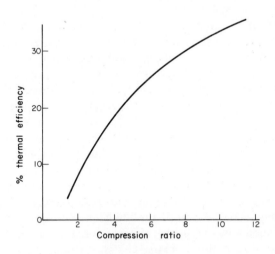

FIG. 25.2. Variation of engine efficiency with compression ratio.

deposit is burned off, and returned in due course to the incoming feed stream and thence back to the cracking chamber. Cracked gases are taken from the top of the cracking chamber directly to a fractionating column. Operating temperatures are typically about 550°C at pressures a little above atmospheric pressure, conditions much milder than those used in the earlier thermal processes.

The 'cat cracker' is a prominent part of the modern refinery. Processes of the type described above can be operated for periods sometimes measured in years before shut-down is necessary, with throughput running up to one million gallons per day.

25.5. Gasoline: quality

Fuel is consumed in a spark ignition engine by igniting a mixture of gasoline and air inside the cylinder head fractionally before the piston reaches the top of its travel. The ensuing combustion should take place rapidly, but smoothly, at flame speeds of about 50 feet per second.

The efficiency of power production in an internal combustion engine is related to the *compression ratio*, i.e. the ratio of the cylinder volume when the piston is at the bottom of its stroke to the cylinder volume when the piston is at the top of its stroke. The variation of E, the efficiency, with r, the compression ratio, is shown in Fig. 25.2. In the last two or three decades compression ratios have steadily increased with the development of more efficient and powerful engines. Before the Second World War, a family saloon car might have a compression ratio of 6·5:1; now values of 8–10:1 are common.

Increased compression ratios, however, have brought problems as well as benefits. A fuel–air mixture can ignite spontaneously on compression, and should this happen in a spark ignition engine a very rapid combustion occurs, producing shock waves of much greater velocity than normal. These waves, reverberating throughout the cylinder head, produce a characteristic sound which gives this preignition phenomenon the name *knocking* or *pinking*. Engine knock results in a drop in efficiency, and can also cause mechanical damage. As we might expect, the higher the compression ratio of an engine, the greater the tendency for knocking to occur.

Gasolines vary in their resistance to knocking. The increase in compression ratios has therefore been accompanied by an increase in gasoline quality with respect to knock resistance. This latter property of a gasoline cannot be defined in absolute terms but is indicated on an arbitrary scale called the *octane rating*.

Octane numbers are simply defined. The hydrocarbon *n*-heptane, C_7H_{16}, which has very poor resistance to knocking, is given an octane number of zero; the hydrocarbon iso-octane (2,2,4-trimethylpentane, C_8H_{18}), which has a high resistance to knocking, is given an octane number of 100. *The octane number of a gasoline is the percentage of iso-octane in a mixture of iso-octane and n-heptane having the same knock resistance as the gasoline under test.* Testing is carried out in a standard engine of variable compression ratio, the ratio being increased gradually until the onset of knocking for the fuel under test. A number of standard engines have been specified, and each need not necessarily give the same result. Octane ratings for motor gasolines may be reported as the octane number (motor method) or the octane number (research method). The latter rating, usually a few points higher for high octane fuels, is most commonly quoted.

25.6. Octane rating and chemical structure

There is a clear relationship between the octane number of a hydrocarbon and its structure. In Table 25.1 we can compare the octane numbers of a series of C_7 hydrocarbons. The progression is such that the octane number rises through the series *n*-alkanes → branched chain alkanes → cycloalkanes → alkenes → aromatics. Multiple branches in an alkane, particularly if the

TABLE 25.1. OCTANE NUMBER (RESEARCH) OF SOME C_7
HYDROCARBONS

Compound	Structure	Octane number
n-Heptane	$CH_3\cdot(CH_2)_5\cdot CH_3$	0
3-Methylhexane	CH_3 $\|$ $CH_3\cdot CH_2\cdot CH\cdot CH_2\cdot CH_2\cdot CH_3$	52·0
3-Ethylpentane	$CH_2\cdot CH_3$ $\|$ $CH_3\cdot CH_2\cdot CH\cdot CH_2\cdot CH_3$	65·0
Methylcyclohexane	CH_3 $\|$ CH $H_2C \quad CH_2$ $H_2C \quad CH_2$ C H_2	74·8
2-Methyl-1-hexene	CH_3 $\|$ $CH_2{=}C\cdot CH_2\cdot CH_2\cdot CH_2\cdot CH_3$	90·4
2,3-Dimethylpentane	$CH_3 \ CH_3$ $\| \quad \|$ $CH_3\cdot CH\cdot CH\cdot CH_2\cdot CH_3$	91·1
2,4-Dimethyl-1-pentene	$CH_3 \quad CH_3$ $\| \quad \|$ $CH_2{=}C\cdot CH_2\cdot CH\cdot CH_3$	99·2
2,2,3-Trimethylbutane	$CH_3 \ CH_3$ $\| \ /$ $CH_3\cdot C\cdot CH\cdot CH_3$ $\|$ CH_3	100 + 1·8†
Toluene	CH_3	100 + 5·8†

†See Section 25.8.

branches are methyl groups, give rise to very good octane ratings, e.g. 2,2,3-trimethylbutane.

Compression ratios are such today that gasolines of octane number above 90 are required if engine knock is to be avoided. Straight-run gasoline from crude petroleum is likely to have an octane number no greater than about 60. Clearly some upgrading procedure is necessary. This can be achieved by blending low rating gasolines with, for example, aromatic hydrocarbons. Much more important, however, are a number of high temperature conversion processes.

25.7. Production of high octane gasoline

We have seen (Section 25.4) that the heavy demand for greater quantities of gasoline has been met by the use of thermal and catalytic cracking processes. The demand for a better quality of gasoline has been met in the same way.

If a gasoline fraction of low octane number is exposed to high temperatures for a short period of time (perhaps 1 to 10 seconds) many of the molecules isomerize rather than break down into smaller fragments. Straight chain alkanes isomerize to branched chain structures, and to cycloalkanes which in turn partially dehydrogenate to aromatics. Simple alkenes are also produced. The information quoted in Table 25.1 indicates that these changes must result in an increase in the octane number of a gasoline so treated.

Thermal and thermo-catalytic processes for upgrading gasoline are called *reforming* processes; a number of these are in widespread use. *Hydroforming* employs a molybdenum catalyst at an elevated pressure of about 250 pounds per square inch and temperature of 500°C. Hydrogen gas is present to saturate alkenes produced by cracking and so to decrease coke formation at the catalyst surface. Cyclization and dehydrogenation of cycloalkanes to aromatics still takes place, however. The closely related and more recent *platforming* process, like hydroforming, is widely applied to gasoline reforming. Both procedures are also used for the synthesis of primary aromatic hydrocarbons (Section 21.2).

Low-temperature processes are also used to obtain structures of high octane rating. Catalytic isomerization of *n*-butane to isobutane followed by cracking of the product to isobutene is used to produce large quantities of the branched C_4 alkene. This fraction can subsequently undergo alkylation (Section 20.4) to produce highly branched high octane-rating hydrocarbons. The product, called *alkylate*, has an octane rating of greater than 90 and is blended with other suitable streams.

Gasoline fuel need not be hydrocarbon material. Alcohols have good octane ratings and methyl and ethyl alcohol are sometimes used as components of a final gasoline blend.

25.8. Additives

Commercial gasolines contain a number of additives, the most important of which is tetraethyl lead, $Pb(CH_2CH_3)_4$.

A number of organometallic compounds help to combat knocking tendencies in a gasoline and therefore effectively raise the octane rating. Tetraethyl lead is not the best of these compounds but it is to date the cheapest. Up to about 4 cm³ per gallon may be added, the effect of the additive varying in different gasolines. Use is made of tetraethyl lead in classifying fuels of higher octane rating than the standard iso-octane. Thus toluene, which is said to have an octane rating of '100 + 5·8', gives identical performance under test conditions to iso-octane containing 5·8 cm³ tetraethyl lead per gallon.

Other additives in a typical gasoline will include anti-oxidants (often amines) to prevent gum formation, rust inhibitors, carburettor de-icing compounds, and a dyestuff.

25.9. Sulphur in gasoline

The presence of sulphur is as undesirable in gasoline as it is in any fuel. Apart from producing acid gases on combustion, sulphur compounds interact with tetraethyl lead rendering it ineffective and so reducing the octane number of the fuel. The fuel is said to have a depressed *lead susceptibility*. Moreover, sulphur present as mercaptans produces really foul smells which do nothing for the market image of the product.

Sulphur compounds can be removed by sulphuric acid treatment, a refining process which can also be used for the removal of alkenes, the precursors of polymeric gums in gasoline. Solvent extraction with, for example, strong alkali is also used to reduce the sulphur content. Improving a gasoline by the removal of mercaptans is termed 'sweetening'; this may entail mercaptan extraction by a process of the type mentioned above, or it may be a process in which the sulphur content is not reduced, but the mercaptans are merely converted to a more innocuous form of sulphur. Hence the *Doctor process* is used to convert mercaptans to disulphides by lead plumbite treatment, carried out, paradoxically, in the presence of a carefully calculated amount of free sulphur:

$$Na_2PbO_2 + 2RSH + S \rightarrow RSSR + PbS + 2NaOH$$

A similar result can be achieved using copper chloride.

25.10. "What petrol should I use?"

Gasoline technology is complex. The procedures we have been considering are well established, and it is reasonable to assume that a customer will find little, if any, difference between different brands of the same grade of petrol. The only real variable wherein the consumer can exercise choice is that of octane rating. In the U.K. a British Standard defining different grades of gasoline according to octane rating was introduced in 1967, and this went a long way towards rationalizing what had been a rather confusing situation. The Standard divides gasolines into four groups, each of which is given a star grading. The scheme is summarized in Table 25.2, and discussed in an independent report.†

†*Which*, April 1967, Consumer Association, London.

TABLE 25.2. BRITISH STANDARD
GASOLINE GRADINGS (B.S. 4040,
part 11, 1967)

Star rating	Research octane number
5 star	100 or greater
4 star	97 or greater
3 star	94 or greater
2 star	90 or greater

An earlier report from the same source‡ emphasized that nothing is to be gained by using a quality of gasoline better than is necessary to prevent knocking in an engine. Indeed, car manufacturers will not recommend five star fuel for medium compression engines, where it can cause damage. The advice of the aforementioned reports should be followed, viz. use the cheapest grade of gasoline which does not cause knocking in your car.

25.11. Diesel fuels

The diesel engine is a compression ignition engine which uses petroleum fuels distilling between about 180°C and 360°C. In the cycle of operations, air is passed into the cylinder head during an upstroke. Compression results in a sharp rise of pressure and temperature, and fuel, sprayed in towards the end of the compression stroke, begins to ignite spontaneously. Fuel feed and ignition continue during the downward power stroke.

The process of injecting oil into a combustion chamber such that it breaks up into very small droplets is called *atomization*. Efficient atomization is necessary in a compression ignition engine (or for that matter in an oil-fired furnace) if combustion is to be complete. For diesel fuels, which are usually selected from the gas oil range of the 'petroleum spectrum', viscosity is an important factor in determining ease of atomization. Diesel oil viscosities range from about 30 Redwood seconds at 100°F up to 100 times that figure, heavier low-speed engines being able to accommodate more viscous oils.

‡*Which,* Jan. 1964, Consumer Association, London.

Carbonaceous deposits are undesirable in diesel engines, as they interfere with the fuel injection system and contaminate the lubricating oil. Here again specifications become less exacting as the engine speed decreases. A high-speed engine may require a light gas oil with a Conradson carbon residue of 0·05 per cent or less, whereas a very low-speed engine may accept an oil of carbon residue in excess of 10 per cent.

25.12.   Cetane numbers

If we contrast the mode of operation of a gasoline engine, where preignition by compression must be avoided, to that of a diesel engine, where compression ignition is central to the whole power cycle, we might well expect logical differences in the fuels most suitable for each type of internal combustion engine.

The *cetane rating* for diesel fuels is equivalent to the octane rating for gasolines, and is a measure of the ease with which a given fuel will undergo compression ignition. Two hydrocarbons are specified as the low and high points on the cetane scale, viz.

1-Methylnaphthalene
Cetane no. 0

$CH_3(CH_2)_{14}CH_3$

n-Cetane
Cetane no. 100

the cetane number of a fuel being the percentage of cetane in a mixture of the two standards giving equivalent performance to the fuel under test. Thus the structural requirements for a diesel fuel are precisely those which have to be avoided in gasolines; straight-chain alkanes like *n*-cetane ignite readily on compression, aromatics like 1-methylnaphthalene are too stable to be good diesel fuels. Cetane numbers in practice range from about 25 to 50, increasing with increasing engine speed.

To avoid the comparatively complex cetane number determination, a related quantity called the *Diesel index* is often used as a test of ignition quality, being defined as

$$\text{Diesel index} = \frac{\text{Aniline point (°F)} \times \text{API Gravity}}{100}$$

The diesel index figure is often, but not always, close to and a little higher than the cetane number.

Additives can be used to raise the cetane number of a diesel fuel. Alkyl nitrates such as ethyl nitrate, $CH_3 \cdot CH_2 \cdot O \cdot NO_2$, are among the compounds suitable for this purpose.

25.13. Kerosine

Kerosine, or 'paraffin', once important because of its illuminating flame, decreased in value until the appearance of the aviation jet engine. Kerosine, or kerosine blended with gasoline, is used as fuel for turbine-driven aircraft, and this fraction is consequently now rated an important product of the petroleum industry.

25.14. Fuel oils

While the light petroleum fractions we have been considering are generally used as sources of heat and power, the term 'fuel oil' is reserved for viscous oils of the gas oil and distillate residue type which are used in furnace combustion as alternatives to coal or gaseous fuels. Liquid fuels lend themselves to automatic control to a greater extent than coal and they have a higher thermal storage capacity than the solid fuel. (Fuel oils have a gross calorific value of about 19,000 Btu per pound, a good-quality coal about 15,000 Btu per pound.)

The most important feature of an oil-fired furnace is the burner or atomizer, through which the fuel passes to the furnace. The ultimate atomizing device is a simple wick from whose surface oil vaporizes and passes into the combustion chamber. Wick burners are very limited in throughput and are normally suitable only for domestic heaters and small load boilers serving domestic central heating units. In industrial furnaces a number of atomizing burners are in use. They may depend on imparting a strong centrifugal force to the oil feed which sprays out into the combustion chamber, breaking up into a mist of very fine droplets; alternatively a pressurized air or steam jet impinging on the oil feed can be used to bring about atomization, the mixture then being passed into the combustion chamber. The viscosity of a fuel oil is important in determining its ease of atomization and also the ease and cost of pumping operations. Fuel oils are classified according to viscosity, and commonly the heavy oils must be heated to a predetermined temperature before they can be satisfactorily handled.

The mechanism of combustion in an oil-fired furnace parallels that of pulverized fuel firing. Each oil droplet entering the flame envelope first loses much volatile matter which burns off in a homogeneous gas phase reaction, leaving as residue a carbon particle (called a cenosphere) which undergoes reaction with oxygen in precisely the same way as a coke residue derived from coal. Both oil droplets and pulverized fuel particles follow a burning time law approximately represented by the equation

$$t_c = F_c \, . \, K d_o^2$$

where t_c is the burning time of the volatile matter of a droplet or particle of initial diameter d_0, F_c is a factor dependent on the amount of excess air present, K a complex burning constant dependent on a number of system variables. The d_o^2 term emphasizes the importance of the atomization process in fuel oil combustion.

Considerations of the sort which have been discussed in Section 24.4, in particular those dealing with air supply and smoke emission, are equally applicable to an oil-fired system. Oil requires less excess air than coal for complete combustion and therefore has the advantage of a higher thermal efficiency.

25.15. Other petroleum products

From the heavy residues of crude petroleum a number of valuable products are extracted. Most important of these, perhaps, are the *lubricating oils* (dealt with separately in Chapter 28). A valuable range of *petroleum waxes* and *greases* are obtained by extensive refining of the back-end fractions. *Bitumen* finds an outlet chiefly as a road-making material, but has many other applications. The ultimate solid residue of petroleum distillation is *petroleum coke*. Its very low ash content makes it a suitable source of electrode carbon, and like coal coke it may also be used as a solid smokeless fuel, although the lack of ash leads to a high rate of corrosion of grate bars unless precautions are taken.

If we include in 'other petroleum products' the full range of petrochemicals, our list becomes almost unending. Many times in earlier chapters petrochemical processes were mentioned. Most organic materials can now be made from petroleum, but the industry is also the source of rather more obscure products. Thus *carbon black*, used mainly as a rubber additive to increase the wear resistance of motor tyres, is made from a variety of petroleum fractions, depending on the process being used. *Ammonia*, very much an inorganic product, is now manufactured from hydrogen obtained by steam reforming of light petroleum fractions, the hydrogen subsequently being reacted with nitrogen (Chapter 13). There is currently no sign that the rate of movement to a petroleum-based chemical industry is likely to decrease.

CHAPTER 26

Chemical Explosives

LIKE traditional fuels, traditional explosives release their energy by chemical reaction. In this chapter we shall concern ourselves with explosives of this class, omitting those which depend on nuclear reactions and which are dealt with separately in Chapter 2.

Fuels and explosives have much in common. In both cases strongly exothermic oxidation reactions serve as energy donors. The essential difference is the *speed* at which oxidation occurs; fuels react at a controlled rate while explosives react almost instantaneously, and violently. (Precisely the same dividing line can be drawn between nuclear fuels and nuclear explosives.)

26.1. Physical characteristics

Two phenomena are particularly associated with explosive reactions, viz.

(i) the release in a very short period of time of a large quantity of energy, mainly as heat;

(ii) a great increase in the volume of products (principally gases) compared to the volume of reactants; e.g. nitroglycerine combustion gives a 10^4 volume increase at 3000°C.

Allowed to take place in a confined space, a reaction with these characteristics will produce gases at high temperatures and pressures, and the subsequent shock waves associated with this high-energy system will cause the disintegration of surrounding material.

Explosives vary in their detailed behaviour on detonation, and a number of specification tests have been established. Of these, two are especially useful in assessing different preparations:

(i) *Explosive strength* is a measure of the energy usefully made available by a detonation. It is quoted in calories per gram ('weight strength') or calories per cm^3 ('bulk strength') and is commonly measured against a standard charge of blasting gelatin. This can be done as in the Trauzl test, in which standard charge and explosive under test are each detonated in a small cavity in a lead block, and the volume increases of the cavities compared.

(ii) *Velocity of detonation* (v.o.d.) is a measure of the velocity of the detonation wave passing through the explosive. It can be determined by optical and electrical techniques, or by comparative testing against a standard detonating fuse. Values are commonly quoted in metres or kilometres per second.

While other tests are significant, these two will tell the experienced operator a great deal about an explosive. The job in hand determines the explosive used. For example, shattering a hard material will require a charge with a high explosive strength and a high v.o.d. Avoiding breakage (e.g. in mining lump coal) will entail choosing a charge with just the opposite criteria.

26.2. The chemical structure of explosives

We have noted that explosive systems must generate both a large amount of heat and a large volume of gas. It is possible to compound chemical mixtures which have these characteristics although the individual components may be innocuous in themselves. The first known explosive, gunpowder or blackpowder, belongs to this category, being a mixture of sodium or potassium nitrate, sulphur and carbon which react together violently to give a complex mixture of products including carbon monoxide, carbon dioxide and nitrogen. The typical modern explosive, however, is a single substance containing the necessary structural features within the one constitution to provide a high-speed, exothermic, gas-forming reaction.

What, in practice, are the necessary structural features? Two requirements may be specified and used to classify explosives on a chemical basis:

(i) There must be present in the molecule at least one chemical bond which can be easily broken. Such bonds may be quite stable in so far as their resistance to chemical decay is concerned. (Indeed, chemical stability is a desirable feature of industrial explosives, and decay inhibitors are often added to improve this property.) These bonds, however, must break readily if energy (usually mechanical, electrical or thermal) is applied; they are bonds which have a low *energy of dissociation*, and their dissociation leads to a very fast disintegration of the molecule as a whole. Comparatively few chemical groups are satisfactory in this respect; inspection shows that most explosives contain nitrogen–nitrogen, nitrogen–oxygen, nitrogen–chlorine, or chlorine–oxygen bonds. There are a number of groups containing such bonds, and some of the more important ones are listed in Table 26.1. Most explosives contain more than one of these groups per molecule, although a single group can in some cases confer explosive properties on a molecule.

(ii) There should be present in an explosive molecule a good proportion of oxygen, if possible sufficient to bring about combustion without recourse to aerial oxygen. The *oxygen balance* of a molecule is used to express the quantity of oxygen internally available in the substance, and is calculated on the basis of carbon being only partially oxidized to carbon monoxide. Nitroglycerin, for example, has a positive oxygen balance and is therefore suitable for mixing with a combustible material deficient in oxygen,

TABLE 26.1. BOND TYPES WITH POTENTIAL EXPLOSIVE
PROPERTIES

Bond type	Group structure	Group name
Nitrogen–oxygen	$-N=O$	Nitroso
	$-N\overset{\nearrow O}{\underset{O}{\diagdown}}$	Nitro
	$-O-N\overset{\nearrow O}{\underset{O}{\diagdown}}$	Nitrate
	$-O-N=C$	Fulminate
Nitrogen–nitrogen	$=N-N=$	—
	$-N=N-$	Diazo
	$\overset{+}{-N}=N\overset{-}{\equiv N}$	Azide
Nitrogen–chlorine	NCl_3	Nitrogen trichloride (and related compounds)
Oxygen–chlorine	$-O-\overset{O}{\underset{O}{\overset{\uparrow}{Cl}}}\!\!\rightarrow\!O$	Perchlorate

$$C_3H_5N_3O_9 \longrightarrow 3CO + 2\cdot5H_2O + 1\cdot5N_2 + 1\cdot75O_2$$

Nitroglycerin   Excess oxygen

while trinitrotoluene has to be treated in just the opposite way
because of a negative oxygen balance.

$$C_7H_5N_3O_6 \longrightarrow 6CO + 1\cdot5N_2 + (C + 5H)$$

Trinitrotoluene   Combustible
                                       balance

26.3. Classification of explosives

Explosives may be divided into primary, secondary and
tertiary classes according to the ease with which they can be
detonated. Sensitivity to detonation varies greatly through a
range of substances, but the main classes are well recognized.

(a) *Primary explosives* are those capable of very ready and powerful detonation, perhaps in extreme cases by the lightest of mechanical shocks. Substances in this class are much too dangerous to be handled in bulk. They are used principally as fillers for detonators, their ignition being used to set off explosion of the main mass of material. Mercury fulminate, $Hg(ONC)_2$, is the classical example of a primary or initiating explosive. More recently lead azide, $Pb(N_3)_2$, has come into widespread use, and one or two other members of this highly sensitive group are produced commercially.

(b) *Secondary explosives* form the real meat of commercial preparations. These comprise the group of compounds known as high explosives. While not so sensitive to detonation as primary explosives, some high explosives are nevertheless very readily detonated. The methods by which they can be made stable for normal handling are outlined in Section 26.4. The majority of modern secondary explosives in widespread use contain either nitrate ester groups or nitro groups.

Nitrate ester explosives are formed by reaction of a suitable alcohol with mixed acid (nitric acid and sulphuric acid).

$$-OH + HO \cdot NO_2 \xrightarrow{H_2SO_4} -O \cdot NO_2 + H_2O$$

The most important industrial explosive, nitroglycerine, is made by this route, as are pentaerythritol tetranitrate (PETN) and the polymeric nitrocellulose.

$$
\begin{array}{ll}
CH_2 \cdot O \cdot NO_2 & CH_2 \cdot O \cdot NO_2 \\
| & | \\
CH \cdot O \cdot NO_2 \qquad O_2N \cdot O \cdot CH_2 & -C-CH_2 \cdot O \cdot NO_2 \\
| & | \\
CH_2 \cdot O \cdot NO_2 & CH_2 \cdot O \cdot NO_2 \\
\text{Nitroglycerine} & \text{Pentaerythritol} \\
& \text{tetranitrate}
\end{array}
$$

Nitroglycerine is a particularly potent material, having a high explosive strength and a high velocity of detonation. It is a liquid with a high sensitivity to detonation, and has to be treated with the utmost caution. The more expensive PETN has an even higher velocity of detonation than nitroglycerine, but cost considerations restrict its use to a few special cases. Nitrocellulose is widely used as the solid combustible phase in commercial mixtures. Other nitrate explosives are in limited use.

$$CH_2 \cdot O \cdot NO_2$$

$$O\!-\!CH$$

$$\sim O\!-\!CH \qquad CH \sim$$

$$CH\!-\!CH$$

$$O_2N \cdot O \qquad O \cdot NO_2$$

Nitrocellulose

Nitro explosives are usually aromatic compounds, formed by the nitration with mixed acid of suitable aromatic rings. The outstanding example in this class is trinitrotoluene (TNT), a solid explosive of medium explosive strength. Another interesting nitro explosive is cyclotrimethylenetrinitramine (RDX) in which the nitro groups are attached to heterocyclic nitrogen atoms. It is a powerful explosive of high velocity of detonation, but is costly.

Trinitrotoluene   Cyclotrimethylenetrinitramine

(c)*Tertiary explosives* are less widely recognized as forming a class in their own right. They are materials which of themselves are not generally regarded as explosives but which in contact with other combustible material can form very dangerous mixtures. The really important member of this class is ammonium nitrate, which decomposes exothermically to give a positive oxygen balance and a high gas content in the products:

$$NH_4 \cdot NO_3 \longrightarrow 2H_2O + N_2 + \tfrac{1}{2}O_2$$

The explosion of ammonium nitrate stores has been the cause of a number of disasters. The characteristics indicated by the above equation make it a valuable constituent of explosive mixtures.

26.4. Commercial preparations

Commercial explosive preparations are, typically, multi-component mixtures containing a high explosive mixed with other material which in itself may or may not be combustible. The use of mixtures permits control not only of explosive strength and velocity of detonation, but also of cost and safety factors. The classification of commercial explosives is most conveniently made according to the high explosive used.

Nitroglycerine-based explosives were first properly developed by Alfred Nobel, a Swedish engineer. Nobel discovered that sensitivity to detonation of the highly dangerous oily liquid could be greatly reduced by absorption on an inert porous material, kieselguhr, which can absorb up to three times its own weight of nitroglycerine. *Dynamite* is such a mixture, although combustible absorbents such as sawdust, and sodium or potassium nitrate, are now commonly added in place of kieselguhr. *Ammonia dynamites* are mixtures in which nitroglycerine has been partially replaced by ammonium nitrate. In the range of nitroglycerine gelatins, such as *Blasting Gelatin* (Nobel's term) or *Gelignite*, the absorbent is nitrocellulose. A nitroglycerine gelatin may be totally organic, and therefore have a high resistance to water, making these very powerful explosives suitable for wet working conditions.

Trinitrotoluene-based powders are widely used in engineering practice. They generally consist of TNT and ammonium nitrate; Amatol is one such mixture. A special class, the Ammonals, contain in addition aluminium powder.

Ammonium nitrate mixed with fuel oil forms a widely used explosive. Neither ingredient is of itself a high explosive and each can therefore be transported in bulk to the place of usage with less stringent safety measures than usual and mixed on the spot. Preprepared commercial mixtures are also available, and are considered to be safer in handling characteristics than other explosives (although ammonium nitrate in bulk should always be treated with caution).

CHAPTER 27

High Polymers

A POLYMER is a substance composed of a number of repeating units, each repeating unit being derived from a starting material called the *monomer*. The linking together of monomer units is called *polymerization*, and if this reaction proceeds to an extensive degree the high molecular weight product is called a *high polymer*.

Polymeric materials occur widely in nature. Thus protein polymers in one form or another are one of the main constructional materials of animal life; carbohydrate polymers, particularly cellulose, serve the same purpose in plant life. Inorganic polymers also exist naturally; many minerals are complex silicate polymers, while diamond and graphite are high molecular weight aggregates of carbon. These are some of the more obvious examples of polymers in nature. There are many others.

Despite the abundance of first-hand examples, it was not until well into the twentieth century that chemists began to understand the true nature of polymers. This understanding has led to a situation where something like a third of the organic chemical industry is now involved in synthetic polymer production. It is difficult to imagine our surroundings without the range of polymeric materials now available, although these materials have emerged in the main only in the last 40 years or so. For most of us this is much more the Polymer Age than the Atomic Age.

370

27.1. Polymerization

Consider the reaction whereby an ester is formed from an alcohol and a carboxylic acid:

$$R \cdot OH + HO \cdot OC \cdot R' \rightarrow R \cdot O \cdot OC \cdot R' + H_2O$$
Alcohol Carboxylic acid   Ester

In the above example, both alcohol and acid have only one reactive functional group, i.e. for each component the functionality number f is one, and together they are said to form an $f(1,1)$ system. The ester product in the above reaction has no free functional groups and the value of f is zero after a one-step reaction.

If we start with an $f(2,1)$ system, i.e. dialcohol and monoacid or diacid and monoalcohol, two steps are required before a diester with zero functionality is obtained:

$$HO \cdot R \cdot OH + HO \cdot OC \cdot R' \rightarrow HO \cdot R \cdot O \cdot OC \cdot R' + H_2O$$
Dialcohol   Monoacid   Half ester
$f = 2$ $f = 1$ $f = 1$

$$HO \cdot R \cdot O \cdot OC \cdot R' + HO \cdot OC \cdot R' \rightarrow R' \cdot CO \cdot O \cdot R \cdot O \cdot OC \cdot R' + H_2O$$
Diester
$f = 1$ $f = 1$ $f = 0$

The feature of these reactions is a decrease in total functionality with each step. If, however, we commence with an $f(2,2)$ system, i.e. diacid and dialcohol, each ester-forming step results in a product which itself has $f = 2$. There is no loss in functionality:

$$HO \cdot R \cdot OH + HO \cdot OC \cdot R' \cdot CO \cdot OH \rightarrow$$
$f = 2$ $f = 2$
$$HO \cdot R \cdot O \cdot OC \cdot R' \cdot CO \cdot OH + H_2O$$
$f = 2$

$$HO \cdot R \cdot O \cdot OC \cdot R' \cdot CO \cdot OH + HO \cdot R \cdot OH \rightarrow$$
$f = 2$ $f = 2$
$$HO \cdot R \cdot O \cdot OC \cdot R' \cdot CO \cdot O \cdot R \cdot OH + H_2O$$
$f = 2$

The reaction, then, can go on indefinitely to form high polymeric material.

This reaction illustrates a general principle, viz. *only f(2,2) systems yield polymeric material.* Reaction in systems of lower functionality ceases at an early stage.

The above example of a polymerization sequence involves condensation reactions in which a water molecule is split out each time alcohol and acid functions react. *Condensation polymers* are those formed by reactions in which some small molecule (e.g. water, ammonia, hydrogen chloride) splits out at each step. The other major class of polymers, *addition polymers* (Section 20.4), are formed by the linking together of initially unsaturated monomer units, usually alkenes or substituted alkenes. No by-product splits out. In addition polymerization the mechanism of the reaction differs from condensation polymerization, but the end result is the same — a large increase in molecular weight occurs. The functionality concept is still valid, for each carbon–carbon double bond, if broken, provides two reactive sites, i.e. an alkene is potentially difunctional:

$$CH_2{=}CH_2 \xrightarrow[\text{or catalyst}]{\text{heat}} {-}CH_2{\cdot}CH_2{-}$$

Ethylene   Difunctional
                                  fragment

$$CH_2{=}CH \xrightarrow{\text{polymerize}} \sim CH_2{\cdot}CH{\cdot}CH_2{\cdot}CH \sim$$
$$\quad\ \ \ \ |\qquad\qquad\qquad\qquad\quad\ |\qquad\ \ |$$
$$\quad\ \ \ \ X\qquad\qquad\qquad\qquad\quad\ X\qquad\ X$$

Substituted   Polyalkene
alkene

27.2. Thermoplastic and thermosetting polymers

The division of polymers into addition and condensation types is useful, but limited. A more practical classification based on the behaviour of the finished product divides a large number of polymers into *thermoplastic* and *thermosetting* materials.

Thermoplastic polymers can be heated to softening without any chemical degradation occurring. Thermosetting polymers on the other hand, once moulded, cannot be melted down without chemical degradation occurring; old material, therefore, cannot be used in new moulding processes.

This difference in behaviour is simply explained in terms of chemical structure. We have seen that an $f(2,2)$ system is the lowest order from which polymer may be obtained. With only two reactive sites available in the growing polymer, such a system can only form long chain-like structures and no chemical linkages will occur between chains. An $f(2,2)$ system is said to form *linear* polymer, and Fig. 27.1 represents an idealized picture of a mass of amorphous linear polymer in which M signifies the repeating unit in the chain. In melting a linear polymer, i.e. in rendering the system fluid such that any one polymer chain can move freely with respect to its neighbour, the only forces which have to be overcome are relatively weak physical forces of attraction between chains. Much chain entanglement will occur, and polar attractions will also be present, but these are easily overcome. More particularly, no chemical bonds need be broken to free one chain from another, and melting is not accompanied by thermal degradation. Linear polymers, then, behave as thermoplastics. The low heats of fusion values of these materials, which are usually between 1 and 12 kcal per mole, reflect the relative ease of the melting process.

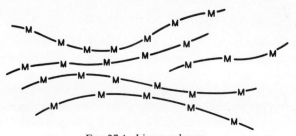

Fig. 27.1. Linear polymer.

If now we consider a system of higher functionality, e.g. an $f(2,3)$ system such as a diacid reacting with a trialcohol, a new factor will be seen to emerge. Let us imagine that initial polymerization yields linear polymer containing the still unreacted third alcohol function.

HO·R·OH + HO·OC·R'·CO·OH ⟶
  |
  OH
                                    ~ O·R·O·OC·R'·CO ~
                                       |
                                       OH

Trialcohol   Diacid   Polymer with free
                                    alcohol function

$f = 3$ $f = 2$ $f = 1$

In this case it is now formally possible that one chain can be linked to another by chemical bonding through further reaction of the free alcohol functions with a diacid molecule.

~ O·R·O·OC·R' ~   ~ O·R·O·OC·R' ~
   | |
   OH   O
   | |
   CO·OH   CO
   | |
   R'   ⟶   R'
   | |
   CO·OH   CO
   | |
   OH   O
   | |
~ O·R·O·OC·R' ~   ~ O·R·O·OC·R' ~

A *cross-linking reaction* of this kind leads to the formation of a three-dimensional rigid polymer network in which the principal chains are firmly linked to one another. Any effort to melt down such material means the breaking of chemical bonds, a high-energy process compared to that of melting thermoplastic polymer, requiring perhaps 90 to 100 kcal per mole. The rupture of chemical bonds on heating would mean the substance being irreversibly degraded and destroyed. Cross-linked systems are therefore described as thermosetting.

Any polymer-forming system which has a functionality of $f(2,3)$ or higher can cross-link. In manufacture, the degree of cross-linking can be varied continuously by varying the amount of trifunctional material present. Hence a gradation in physical properties from a highly thermoplastic material to a highly cross-linked thermosetting material can be achieved.

27.3. Molecular weight distribution in polymers

Organic monomers, often gases or low boiling liquids, can be polymerized to useful constructional materials. The great increase in molecular weight which the reaction produces is the key to the usefulness of the process. The experimentally determined molecular weight of a polymer sample is an average value, for polymerization yields a product which is a mixture of large molecules, each of essentially the same chemical structure, but of varying size and molecular weight. In a typical thermoplastic with an average molecular weight of 100,000 there may well be individual chains with molecular weights as low as 5000 or as high as 2,000,000. The physical properties of a polymer vary with its average molecular weight, which must therefore be controlled. For example, polyacrylonitrile, used as a synthetic fibre, does not exhibit fibre-forming characteristics until a molecular weight of between 10,000 and 20,000 is achieved. Again, the glass transition temperature of thermoplastics, discussed in the following section, rises rapidly with increasing molecular weight before approaching a final steady value.

27.4. Physical properties of polymers

Polymers differ markedly in physical behaviour from other more traditional materials of construction. They are not so susceptible to certain modes of weathering and corrosion, and so are used for making, for example, pipes and drains, gear wheels, perhaps in the future car bodies. They are light and easily moulded, and so permit the possibility of large-scale prefabrication in the building industry. Being essentially covalent, they have low thermal and electrical conductivities, and this fact combined with their low water-retention characteristics makes them excellent insulating materials. They can be made with a lack of rigidity extremely useful in kitchen utensils and toys. Examples of the wide use of polymers are not difficult to find. The major disadvantage of organic polymers is that of carbon compounds in general; they do not respond well to high temteratures, at which they will undergo either combustion or thermal degradation. Even this drawback has been partly overcome by the development of new organic and, more particularly, inorganic polymers.

Linear and partially cross-linked materials can exist in a number of different physical states which in turn affect their bulk properties. At sufficiently low temperatures many plastics exist in the solid phase as amorphous inflexible aggregates called *glasses*. If a polymeric glass is heated a fairly well-defined temperature called the *glass transition temperature*, T_g, is observed at which the polymer chains become more flexible and the material is converted to a *rubber*, i.e. a material capable of undergoing considerable reversible elastic deformation. As a polymer passes through the glass transition temperature bulk properties such as the specific heat or the modulus of elasticity undergo change, and it is by following the change in such a property or in its temperature coefficient that the glass transition temperature is measured. The glass transition temperature, which is not an absolute value, varies widely for a range of polymers. For example, polyethylene has a reported value of $-122°C$, polyvinyl acetate $28°C$, polyvinyl chloride $70°C$, and polystyrene $94°C$. Glass transition values can be altered by, for example, altering the method of polymerization, or by adding *plasticizers* to the polymer. Plasticizers, which are bulky molecules like dibutyl phthalate, act as internal lubricants in a polymer, inserting themselves between the polymer chains and so weakening the cohesive forces which hold them together in

$$CO \cdot O \cdot CH_2 \cdot CH_2 \cdot CH_2 \cdot CH_3$$

$$CO \cdot O \cdot CH_2 \cdot CH_2 \cdot CH_2 \cdot CH_3$$

Dibutyl phthalate

the solid state. The result is that the glass transition temperature is lowered, and the material becomes more flexible. Loss of plasticizer can occur by accidental solvent extraction, or perhaps by evaporation in continually warm conditions, and it is often made evident by an obvious increase in brittleness in the bulk material.

In the glass and rubber states polymers are amorphous, but they can also exist in a crystalline state. Crystallinity can be equated with order, and in a polymer can vary from zero in a totally amorphous system to almost 100 per cent in a highly ordered system. The appearance of both crystalline and non-crystalline regions in a polymer mass is represented in Fig. 27.2. We shall see (Sec. 27.6) that the degree of crystallinity in a polymer can vary with the method of initiating the reaction. It can also be encouraged by process techniques such as annealing or cold drawing, or, as in the case of polyamides (Section 27.8), by having present in the polymer chemical structures which have an inherent tendency to arrange themselves in

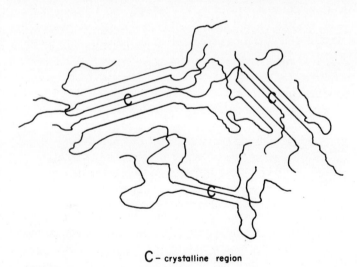

C – crystalline region

FIG. 27.2. Crystalline and non-crystalline regions in a polymer mass.

ordered patterns. Highly crystalline polymers exhibit a well-defined *crystalline melting point*, T_m. With increasing crystallinity in a polymer there is a corresponding decrease in elasticity, and polymer crystallinity is therefore an important parameter used in controlling the mechanical properties of the finished material.

The interrelations between the glassy state, the rubber state and the crystalline state in a polymer cease to apply in a highly cross-linked system. Progressive cross-linking leads to progressive rigidity, and a fully thermosetting rigid material is likely to have poorer elastic deformation characteristics than, for example, steel. Such a material, however, while fully cross-linked may still be amorphous.

27.5. Cellulose plastics

Cellulose is a naturally occurring polymer of complex structure, forming the major part of vegetable cell wall structures. The raw material, which is extracted from sources such as cotton, wood pulp and straw, can be processed to give a range of useful plastics. Inspection of a segment of the cellulose

Partial structure of cellulose

chain shows that each ring in the chain contains three alcohol hydroxyl groups. Cellulose usually undergoes chemical modification to improve its characteristics in various directions by carrying out one of a number of reactions at these hydroxyl functions.

Cellulose may be nitrated by reaction with mixed acid. The extent of nitration can be varied, so varying the properties

$$-OH + HNO_3/H_2SO_4 \longrightarrow -O \cdot NO_2 + H_2O/H_2SO_4$$

of the product. *Cellulose nitrate*, mixed with camphor as plasticizer, is marketed as a moulded plastic under such names as Celluloid or Xylonite. It tends to discolour in time, and is highly inflammable. Indeed fully nitrated cotton cellulose, gun-cotton, is a principal ingredient of many industrial explosives.

Reaction of cellulose with acetic acid yields the ester *cellulose acetate*, less inflammable than the nitrate and suitable

$$-OH + HO_2C \cdot CH_3 \longrightarrow -O \cdot OC \cdot CH_3 + H_2O$$
Acetic acid

for moulding or film castings; partially acetylated cellulose is used as the base for photographic film. Fully acetylated cellulose is suitable for fibre production and appears under such names as Tricel.

Regenerated cellulose is the name given to a cellulose which has been extracted, chemically treated, and then reconverted to be used as cellulose rather than as one of its derivatives. Regeneration processes usually cause a drop in the molecular weight of the product. The proprietary transparent sheeting Cellophane is a regenerated cellulose; so too is most rayon, or artificial silk.

27.6. Vinyl and related thermoplastics

Thermoplastics formed by addition polymerization account for a very large and important fraction of polymer production. Within this field there exists the sub-group of *vinyl polymers*, which are obtained from monomers of general formula $CH_2{=\!=}CH \cdot X$ whose structures include the vinyl $CH_2{=\!=}CH-$ radical. Most of the principal thermoplastics are vinyl polymers, and with their monomers these are listed in Table 27.1.

TABLE 27.1. VINYL MONOMERS AND POLYMERS

Monomer	Structure	Polymer
Ethylene	$CH_2{=}CH_2$	Polythene
Propylene	$CH_2{=}CH$	Polypropylene
	$\quad\quad\mid$	
	$\quad\quad CH_3$	
Vinyl chloride	$CH_2{=}CH$	Polyvinyl chloride (PVC)
	$\quad\quad\mid$	
	$\quad\quad Cl$	
Vinyl acetate	$CH_2{=}CH$	Polyvinyl acetate (PVA)
	$\quad\quad\mid$	
	$\quad\quad O{\cdot}CO{\cdot}CH_3$	
Styrene	$CH_2{=}CH$	Polystyrene
Acrylonitrile	$CH_2{=}CH$	Polyacrylonitrile
	$\quad\quad\mid$	
	$\quad\quad CN$	

Some thermoplastics are also derived from unsaturated monomers other than vinyl compounds, and three of the more significant examples are listed in Table 27.2.

TABLE 27.2. NON-VINYL MONOMERS AND POLYMERS

Monomer	Structure	Polymer
Vinylidene chloride	$CH_2{=}CCl_2$	Polyvinylidene chloride
Methyl methacrylate	$CH_2{=}C{\cdot}CO{\cdot}O{\cdot}CH_3$	Polymethyl methacrylate
	$\quad\quad\mid$	
	$\quad\quad CH_3$	
Tetrafluorethylene	$CF_2{=}CF_2$	Polytetrafluorethylene (PTFE)

There are a number of ways in which unsaturated monomers can be induced to polymerize. Generally some sort of initiator is used. Alkene polymerization has already been described briefly (Section 20.4) and the three-step process of initiation,

Initiation

$$B— + CH_2=CH \longrightarrow B—CH_2 \cdot CH—$$
$$\qquad\qquad\quad | \qquad\qquad\quad\quad |$$
$$\qquad\qquad\quad X \qquad\qquad\quad\quad X$$

Initiation fragment   Monomer

Propagation

$$B—CH_2 \cdot CH— + n \cdot CH_2 = CH \longrightarrow B—CH_2 \cdot CH \cdot (CH_2CH)_n$$
$$\quad\quad\quad |\qquad\qquad\qquad |\qquad\qquad\qquad\quad |\qquad\quad |$$
$$\quad\quad\quad X\qquad\qquad\qquad X\qquad\qquad\qquad\quad X\qquad\quad X$$

Termination

$$B—CH_2 \cdot CH \cdot (CH_2 \cdot CH)_n + R— \longrightarrow B(CH_2 \cdot CH) \cdot R_{n+1}$$
$$\quad\quad\quad |\qquad\quad |\qquad\qquad\qquad\qquad\qquad |$$
$$\quad\quad\quad X\qquad\quad X\qquad\qquad\qquad\qquad\qquad X$$

Growing chain   Reactive   Dead polymer
                  fragment

propagation and termination is summarized above. The final termination step (which in part can be used to control the final molecular weight) is envisaged as a growing chain reacting with another fragment R- which itself may be a second growing chain. Other termination mechanisms are possible. From beginning to end a very short time — perhaps one-thousandth of a second — is required for any one chain to be formed.

The initiating system used can affect quite profoundly the properties of the final polymer. Many vinyl polymers have bulky side groups attached to the backbone, and these tend to interfere with the chain packing efficiency. If polymerization is initiated by generating free radicals† in the presence of monomer, an initiating technique commonly used, the side groups of

†When a chemical bond is broken symmetrically the atomic nuclei concerned each contain a single unpaired electron in the outer shell. Such particles, called *free radicals*, are highly reactive and play important roles in a number of industrial processes; e.g. the cracking of ethane can yield two methyl free radicals:

$$H_3C—CH_3 \longrightarrow 2H_3C \cdot$$

the product are arranged quite haphazardly along each chain. We can see that this must lead to interference of one group with another at many points, and therefore poor packing results. This state of affairs is represented in Fig. 27.3, which it should be noted is a gross distortion of the true shape of the main chain in which each carbon atom has four bonds tetrahedrally distributed about it.

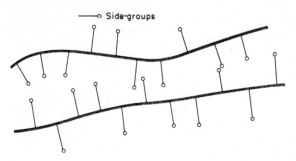

FIG. 27.3. Interfering side-groups in free-radical initiated polymer.

If, instead of using a free radical system, polymerization is initiated by any one of a variety of partially inorganic ionic initiators, e.g. a mixture of aluminium triethyl, $Al(CH_2 \cdot CH_3)_3$, and titanium tetrachloride, $TiCl_4$, chain growth takes place in a controlled manner such that a regular arrangement of side groups about the polymer backbone results. This permits far better packing, as Fig. 27.4, which represents side groups falling on alternate sides of the main chain, indicates.

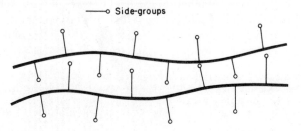

FIG. 27.4. Efficient packing in stereoregular polymer.

Polymers produced from ionic initiator systems are called *stereoregular polymers*; they reflect better packing characteristics by displaying higher degrees of crystallinity and higher densities than free radical initiated products. In the extreme case of propylene, the free radical initiated material is a gum with no useful properties, so great is the interference effect of the bulky methyl side groups; stereoregular polypropylene, a crystalline solid, is of considerable commercial significance. The Italian chemist Natta and the German Ziegler, the two men most closely connected with developments in stereoregular polymerization, were awarded jointly the Nobel prize in chemistry in 1963 for their work, so important is it judged.

Copolymerization is another useful and widely applied method of controlling polymer properties. Here two or more unsaturated monomers are polymerized together to give a mixed polymer whose properties depend on the initial monomer ratios used. Thus styrene and acrylonitrile can be co-polymerized:

Styrene   Acrylonitrile   A–S Copolymer

A short summary of some of the more important thermoplastics is given below. Much more comprehensive information about their properties will be found elsewhere.†

Polyethylene

Polyethylene, or polythene, was one of the earliest vinyl polymers to be marketed (in 1937 in the U.K.). The early I.C.I. process used very high pressures and elevated temperatures to polymerize the gaseous monomer, but newer processes

†For example, *British Plastics Yearbook*, Iliffe, London.

employing Ziegler–Natta type initiators have greatly reduced the intensity of reaction conditions, and almost eliminated the formation of chain branching which can arise due to side reactions. Low density polythene (sp.gr. 0·91–0·93) produced by the earlier processes begins to soften at about 85°C; high density polythene (sp.gr. *ca.* 0·935–0·96) shows a crystalline melting point at about 110°C, and sometimes considerably higher. The high density material, therefore, can be used for carrying hot water where the former may fail.

Polypropylene

Stereoregular polypropylene has qualities which make it suitable for synthetic fibre production; with, in addition, low water-retention figures, polypropylene is being used for rope production as well as being an alternative to polythene.

Polyvinyl chloride

Vinyl chloride polymerizes to a material which is less inflammable than the related hydrocarbon polymers. PVC is now widely used for clothing. As a substitute for shellac, it has made the production of long-playing records possible. It is often produced in modified form by copolymerization with, for example, vinyl acetate or vinylidene chloride.

Polystyrene

Styrene is produced from petroleum sources by the synthesis and subsequent cracking of ethyl benzene.

| Benzene | Ethylene | | Ethylbenzene | Styrene |

This monomer reacts to give one of the very important vinyl polymers. The inflexibility of the benzene rings imparts a brittle quality to the product, which is transparent, and polystyrenes do not exhibit a glass transition until temperatures of about 90°C are reached. Both electrical and thermal resistances are high; expanded polystyrene is now much used as a thermal insulator in the form of tile or granule. Copolymerized with acrylonitrile and other monomers, it forms rubber-like products (Section 27.9).

Polymethylmethacrylate

This polymer, which first appeared in 1933, is marketed as, for example, Perspex (U.K.) and Lucite (U.S.A.). The monomer, methyl methacrylate, is a derivative of acrylic acid, the parent structure to which all acrylic polymers are related:

$$CH_2 {=} CH \cdot CO_2H \qquad CH_2 {=} CH \cdot CO \cdot O \cdot CH_3$$

Acrylic acid Methyl acrylate

$$CH_2 {=} \overset{\overset{\displaystyle CH_3}{|}}{C} \cdot CO \cdot O \cdot CH_3$$

Methyl methacrylate

Methyl methacrylate monomer is obtained by a multistep synthesis from acetone. The polymer has outstanding optical properties, and is widely used as a glass substitute.

Polytetrafluoroethylene

Fluorine-containing polymers are expensive, and PTFE, the simplest and best known of the group, is no exception. Sold under the trade names of Teflon and Fluon, this high-density material (sp.gr. *ca.* 2·2) has some quite outstanding properties: a softening temperature of almost 400°C, negligible water-retention figures, a very low coefficient of friction which gives it special lubricating qualities, and a high degree of chemical stability.

27.7. Thermosetting resins

The principal thermosetting resins are condensation poly-mers. In the U.K. currently the output of thermosetting resins is running at about 300,000 tons per annum, i.e. approximately half that of thermoplastics. Finished products in this field often contain a large percentage of inert filler, e.g. sawdust or asbestos fibre in a moulded article, paper or cloth sheets in a laminate.

Phenol-formaldehyde resins

The reaction between phenol and formaldehyde (Section 22.6) introduces one or more methylol —CH_2OH groups into the benzene ring:

Phenol   Formaldehyde   Methylolated phenol

These highly reactive groups can undergo further condensa-tion reactions with one another to produce ether bridges and methylene —CH_2— bridges between the rings.

$$—CH_2OH + HOCH_2— \longrightarrow —CH_2 \cdot O \cdot CH_2— + H_2O$$
Ether bridge

$$—CH_2 \cdot O \cdot CH_2— \xrightarrow{heat} —CH_2— + CH_2O$$
Methylene   Formaldehyde
bridge

Each benzene ring has a potential functionality of three, and so extensive cross-linking can be achieved. If curing were taken to completion, a three-dimensional network, containing only methylene bridges in the ideal case, would be obtained.

By ensuring a deficiency of formaldehyde in the first stage of manufacture, i.e. by decreasing the average number of methylol groups per ring such that f is two or less in many rings,

Idealized representation of cured phenol-formaldehyde resin

linear thermoplastic products, called novolaks, can be obtained. Later addition of a source of formaldehyde followed by curing will give the fully cross-linked resin. (These and other resins are often supplied as Tube A, the thermoplastic intermediate, and Tube B, the cross-linking reagent. On mixing, the final fully cured product is obtained.)

Phenol-formaldehyde resins were the first totally synthetic polymers to be produced on any scale, appearing in the patent literature in the few years before the First World War. Their development was due to Henri Baekeland, whose name continues to be closely associated with these products in the proprietary name Bakelite.

Amine-formaldehyde resins

Curing reactions in the phenol-formaldehyde system involve a number of side reactions in addition to the cross-linking steps indicated above. One result of these minor reactions is to produce a dark-coloured product. To overcome this problem a variety of clear products based on amine-formaldehyde reactions have been developed.

The initial reaction between formaldehyde and an amine group is two-stage:

$$R \cdot NH_2 + 2CH_2O \longrightarrow R \cdot NH \cdot CH_2OH \longrightarrow R \cdot N(CH_2OH)_2$$

| Primary amine | Formaldehyde | Monomethylol amine | Dimethylol amine |

These methylolated amines can subsequently polymerize to give thermoplastic and thermosetting polymers. In part the condensation reactions are similar to the scheme set out for phenol-formaldehyde polycondensations:

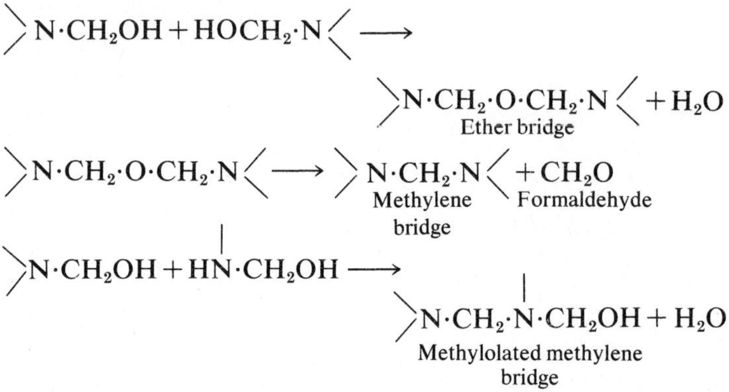

$$>N \cdot CH_2OH + HOCH_2 \cdot N< \longrightarrow$$

$$>N \cdot CH_2 \cdot O \cdot CH_2 \cdot N< + H_2O$$
Ether bridge

$$>N \cdot CH_2 \cdot O \cdot CH_2 \cdot N< \longrightarrow > N \cdot CH_2 \cdot N< + CH_2O$$
Methylene   Formaldehyde
bridge

$$>N \cdot CH_2OH + HN \cdot CH_2OH \longrightarrow$$

$$>N \cdot CH_2 \cdot N \cdot CH_2OH + H_2O$$
Methylolated methylene
bridge

Curing, however, must be much more complex than the above scheme would indicate, for some monomethylol amine derivatives, with an apparent functionality f of one, can give linear polymers, and some dimethylol derivatives ($f = 2$?), thermosetting resins.

The most important amine components used in industrial polycondensate production are urea and melamine, the latter heterocyclic compound offering the possibility of introducing six methylol groups per ring. Amine-based resins such as Formica and Melmex often contain both urea and melamine units, and other amine components can also be used to modify

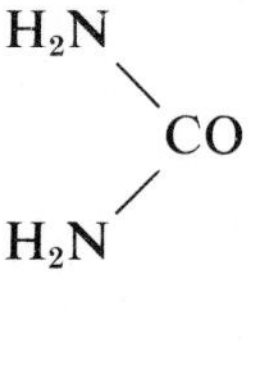

Urea

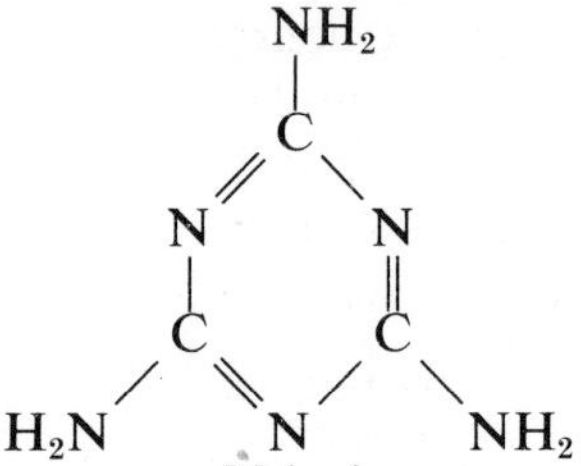

Melamine

the properties of the final resin. These clear resins can readily incorporate light pastel-coloured pigments, either in mould or laminate form. Their stability to light and heat, and their resistance to dirt, solvents, and normal wear and tear, is excellent, the more so when melamine is used in high proportion.

Polyester resins

Polyesters have a number of uses. The class includes a series of materials chiefly employed as protective coatings and adhesives.

Alkyd resins are polyester structures formed from dibasic carboxylic acids (e.g. phthalic acid) and the trifunctional alcohol glycerol.

Phthalic acid ($f = 2$)   Glycerol ($f = 3$)

Alkyd resins are principal components of protective surface coatings of various kinds (e.g. paints, lacquers, enamels) and have good adhesive properties.

Epoxy resins are more difficult to classify chemically; a number are in part polyesters, cross-linked resins being formed by reacting phthalic or other dibasic acids with linear polymers containing free alcohol functions in addition to the characteristic epoxy $>C-C<$ group.

Linear epoxy polymer   Dibasic acid

⟶ Epoxy resin
Thermosetting polymer

Their usefulness in part overlaps with the alkyd resins, epoxy resins having the advantage of even better adhesive properties and a flexibility which, in a coating, means high resistance to cracking. They are also used in laminates and mouldings, appearing in their various forms under trade names such as Araldite and Epon.

Polyurethanes are linear polymers which can be cross-linked by reaction with polyesters to give protective coatings of very high stability and wear resistance. They are expensive, but are now being used where a very high class of finish to an article is justified.

27.8.   Synthetic fibres

It will be appreciated that any one polymer may well be versatile, and may be used as a thermoplastic, or as thermo-setting moulding material, or as a highly adhesive coating, and so on. We are merely emphasizing the uses to which particular polymers are best suited.

For a polymer to make a good synthetic fibre, it must crys-tallize to give a large proportion of chains lying in order paral-lel to one another, and having a high tensile strength in the longitudinal axis. Stereoregular polypropylene has the neces-sary characteristics to make good fibre; so also does poly-acrylonitrile (e.g. Orlon fibre). The two outstanding synthetic fibres, however, are condensation polymers.

Nylons

Nylon is the name given to synthetic polyamides, in which the recurring amide —CO·NH— linkage is formed from a diamine and a dibasic carboxylic acid.

$$HO_2C·R·CO_2H + H_2N·R'·NH_2 \longrightarrow$$
Dibasic acid   Diamine

$$\sim OC·R·CO·NH·R'·NH\sim + H_2O$$
Polyamide *or* Nylon

When the American chemist W. H. Carothers chose to examine polyamides as potential fibre-forming substances he chose wisely, for he was reflecting the practice of nature (animal hair, for example, is a complex polyamide or protein). These structures form good fibres because there exists the possibility of hydrogen bonding between a strongly polarized oxygen atom of one chain and an amide group hydrogen of another chain. It will be seen from Fig. 27.5 that this effect, repeated many hundreds of times along any one chain length,

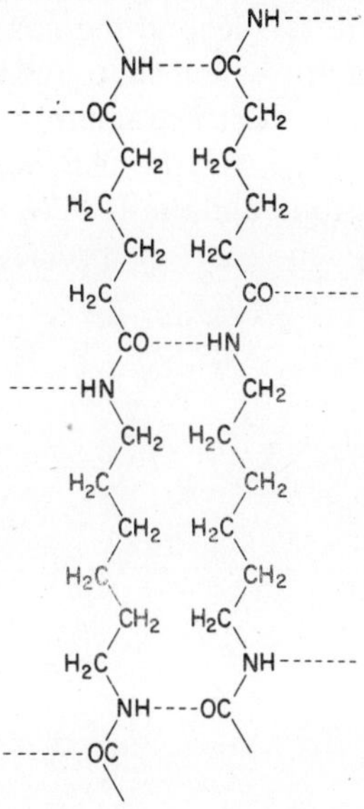

Fig. 27.5. Hydrogen bonding in polyamides.

must assist greatly in aligning the chains in order and producing fibrous, crystalline polymer. While the hydrogen bond effect is very important in polyamides, the way in which nylons and other synthetic fibres are spun and drawn also greatly affects the extent to which crystallinity occurs.

Nylons have a terminology all of their own. The most common nylon in the U.K. and the U.S.A. is that made from adipic acid and hexamethylene diamine. The product is called

$$HO_2C \cdot (CH_2)_4 \cdot CO_2H + H_2N \cdot (CH_2)_6 \cdot NH_2 \longrightarrow$$

Adipic acid   Hexamethylene
diamine

$$\sim OC \cdot (CH_2)_4 \cdot CO \cdot HN \cdot (CH_2)_6 \cdot NH \sim$$

6,6-Nylon

6,6-Nylon, the first digit indicating the number of carbon atoms in the amine segment, the second the number in the acid segment. Thus another nylon which has found favour in the U.S.A., 6,10-Nylon, is formed from hexamethylene diamine and sebacic acid, $HO_2C \cdot (CH_2)_8 \cdot CO_2H$. An interesting nylon now produced both in Europe and the U.S.A. is 6-Nylon, made by polymerization of a cyclic amide, caprolactam.

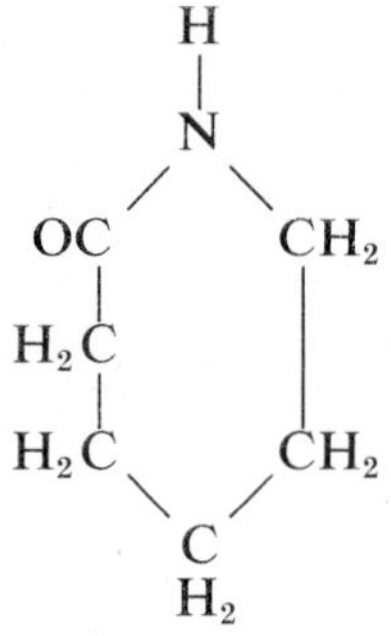

Caprolactam $\xrightarrow{\text{polymerize}}$ $\sim NH \cdot CO \cdot (CH_2)_5 \sim$

6-Nylon

These three nylons dominate the market, the 6,6-version being the most important. The starting point for commercial production of 6,6-Nylon and 6-Nylon, rather surprisingly, is benzene (or phenol), from which may be made all the intermediates.

Nylons form excellent fibres, but their ability to retain physical dimensions after fabrication have also made them popular thermoplastic moulding materials.

Polyethylene terephthalate

Polyethylene terephthalate fibre is marketed as Terylene (U.K.) and Dacron (U.S.A.); it also appears as Melinex film and sheeting. It is a polyester, formed from terephthalic acid and ethylene glycol.

$$HO_2C-\underset{\text{Terephthalic acid}}{\underbrace{\bigcirc}}-CO_2H \;+\; \underset{\text{Ethylene glycol}}{HO\cdot CH_2\cdot CH_2\cdot OH}\longrightarrow$$

$$OC-\bigcirc-CO\cdot O\cdot CH_2\cdot CH_2\cdot O \;+\; H_2O$$

Polyethylene terephthalate

The polymer owes its fibre-forming qualities to the presence of recurring planar benzene rings. In a polymer chain the planar rings act as stiffeners along the chain length. This degree of rigidity in the longitudinal axis is the key structural feature of the polyester fibre. The importance of the linear structure can be emphasized by replacing terephthalic acid by phthalic acid, in which the two acid groups are *ortho*- to one another. With ethylene glycol, this yields a polymeric gum with no fibre forming properties. The shapes of the chains are such that they cannot lie side by side, and the degree of crystallinity is reduced to negligible proportions.

27.9. Rubbers, natural and synthetic

A rubber is a substance with a low modulus of elasticity, i.e. it can be stretched considerably and still return to its original dimensions.

Natural rubber, which is obtained as a milk white latex from the rubber tree, is a polymer of the diene hydrocarbon isoprene (Section 20.5). In the main the isoprene units are linked 'head-to-tail', a residual double bond occurring in the middle of each segment and within the main backbone. Other arrangements of the isoprene unit in which the double bond is external to the main chain make minor contributions to the overall structure of rubber latex.

$$\sim CH_2 - \underset{\underset{CH_3}{|}}{C} = CH - CH_2 - CH_2 - \underset{\underset{CH_3}{|}}{C} = CH - CH_2 \sim$$

Segment of rubber chain showing 'head-to-tail' arrangement

$$\sim CH_2 - \underset{\underset{CH_3}{|}}{C} = CH - CH_2 - CH_2 - \underset{\underset{\underset{H_3C}{\diagdown}\overset{}{C}\overset{\diagup}{}\diagup CH_2}{|}}{CH} \sim$$

Segment of rubber chain with external double bond

Polymer of the first structure would be called 1,4-polyisoprene, indicating the 'head-to-tail' linking of monomer units which involves carbon atoms one and four in each isoprene molecule. This name, however, is incomplete. When a non-symmetrical alkene double bond is present in a compound stereoisomers† called *cis-* and *trans-*isomers can occur, depending on the orientation of different groups about the double bond. In the polyisoprene molecule *cis-*isomerism occurs when all methyl groups lie on the same side of the chain, *trans-*isomerism when they alternate from one side to the other. These two quite distinct forms of the polymeric hydrocarbon, *cis-*1,4-polyisoprene and *trans-*1,4-polyisoprene, are represented below:

$$\underset{-CH_2}{\overset{CH_3\quad H}{\underset{\diagup}{C}}} = \underset{CH_2 - CH_2}{\overset{}{\underset{\diagdown}{C}}} \qquad \underset{}{\overset{CH_3\quad H}{\underset{\diagup}{C}}} = \underset{CH_2-}{\overset{}{\underset{\diagdown}{C}}}$$

(a) *cis-*1,4-Polyisoprene

†When two compounds have the same molecular formula but differ from one another in the way in which their atoms are arranged in space, they are called *stereoisomers.*

$$\underset{-CH_2 \qquad\quad H}{\overset{CH_3 \qquad\qquad CH_2-CH_2 \qquad\quad H \quad CH_2-}{C=C \qquad\qquad C=C}}$$

(b) *trans*-1,4-Polyisoprene

In a rubber, individual polymer chains must have space enough to uncoil in response to external forces. The *cis* configuration in polyisoprene packs inefficiently and hence is equated with elastic behaviour. Natural rubber is about 95 per cent *cis*-1,4-polyisoprene. The *trans* isomer, in which much more efficient packing can be achieved, also occurs naturally. It is called gutta percha, a hard white non-elastic substance used for the outside cover of golf balls. Both *cis*- and *trans*-polyisoprene are now produced synthetically from the monomer by the use of catalysts which ensure a stereoregular polymerization.

Natural rubber in its raw state is soft and sticky. It was of very limited use until Samuel Goodyear discovered the process of *vulcanization*. This process is now recognized as a cross-linking reaction between main chains which involves the residual double bonds in the polyisoprene molecule. The cross-links act as internal stiffeners in the polymer. Goodyear

$$\underset{}{\overset{H_3C}{\sim CH_2-C=CH-CH_2\sim}}$$

$$+\,S \longrightarrow$$

$$\underset{H_3C}{\sim CH_2-C=CH-CH_2\sim}$$

Rubber latex

$$\underset{}{\overset{H_3C \quad S}{\sim CH_2-\underset{|}{C}-\underset{|}{C}H-CH_2\sim}}$$
$$\underset{}{\overset{|}{S}}$$
$$\underset{}{\overset{|}{S}}$$
$$\underset{H_3C \quad S}{\sim CH_2-\underset{|}{C}-CH-CH_2\sim}$$

Vulcanized rubber

vulcanized rubber by heating it with sulphur. This produces sulphur bridges between the principal chains. Other crosslinking reagents are now used for vulcanization, but sulphur remains the important one.

Variation in the amount of sulphur added varies the stiffening effect, and rubber can be produced as a highly elastic material on the one hand, or as a highly cross-linked rigid material (ebonite or vulcanite) on the other.

Synthetic rubbers are increasingly taking the place of the natural product, which is stable neither in price nor continuity of supply. A number of monomers can be polymerized to elastic materials, among which not surprisingly isoprene and other dienes are prominent. A summary of some synthetic rubbers is given in Table 27.3, from which it will be seen that copolymers occupy an important place in the industry. This list is by no means exclusive, and a number of synthetic elastomers which include elements other than carbon in their backbone are of some significance — the sulphur-containing Thiokols, which have outstanding solvent resistance; polyurethane

TABLE 27.3. SYNTHETIC RUBBERS

Monomer(s)	Polymer	Characteristics
Isoprene	*cis*-Polyisoprene	Slightly inferior to the natural product
Butadiene	*cis*-Polybutadiene	Retains rubber state at low temperatures; low energy loss material
Butadiene + Styrene	Buna S, SBR, GR–S	Most important synthetic. High abrasion resistance; low energy losses; used for car tyres, etc.
Butadiene + Acrylonitrile	Buna N, NBR	High resistance to hydrocarbon solvents
Chloroprene	Neoprene	High resistance to solvents; to heat; and to oxidation
Isobutylene	Butyl rubber	Highly resistant to gaseous diffusion; resistant to oxidation
Ethylene + Propylene	Ethylene-propylene rubber	Low density rubber similar to SBR rubbers

elastomers for rubber foam production; the inorganic silicone rubbers which withstand high temperatures to a far greater degree than the organic materials; and many others which are often related to the principal synthetics listed in Table 27.3.

The low energy losses noted as particularly characteristic of the butadiene and butadiene-styrene rubbers are worth some consideration. If a piece of rubber is stretched, and then allowed to contract again, a plot of the force applied against the deflection shows that stretching and contraction do not follow identical paths (Fig. 27.6) but form a hysteresis loop. The work

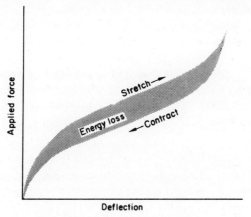

FIG. 27.6. Hysteresis loss in rubbers.

done in each step is given by the area under each curve, and the work or energy loss over the whole cycle is therefore given by the area enclosed by the curves, the so-called hysteresis loss. This loss is easily demonstrated, for it appears as heat, and rubber continually expanded and contracted soon rises in temperature. In car tyres, a low hysteresis-loss rubber is essential, for energy loss here means not only a rise in tyre temperature but also an increase in petrol consumption. Butadiene-styrene rubbers are particularly suitable for tyre manufacture, having a high ratio of energy reabsorbed to energy expended ('high-resilience' rubbers) in the expansion–contraction cycle.

27.10.   Inorganic polymers

Any substance is limited in use by the strength of the chemical bonds existing throughout its mass. Inorganic polymers offer the promise of greater thermal and chemical stability than many of their organic counterparts, whose non-polar carbon–carbon bonds and weakly polar carbon–hydrogen bonds are too readily subject to degradation. *Silicone polymers*, in which the backbone is a chain of alternating silicon and oxygen atoms, have been extensively developed, and are available as liquid and solid lubricants, as rubbers, as resins, and so on. Each silicon atom carries organic side chains, commonly methyl groups, and this combined with the intrinsic strength (or weakness) of

$$-\text{O}-\overset{|}{\underset{|}{\text{Si}}}-\text{O}-\overset{|}{\underset{|}{\text{Si}}}-\text{O}-\overset{|}{\underset{|}{\text{Si}}}-\text{O}-$$

Backbone of silicone polymer

the backbone limits these polymers to a temperature ceiling of about 250°C, much better than most organic polymers, but rather less than many uses require. The technological development of other already known inorganic polymers (and some highly specialized organic polymers) which is already underway promises to raise ceiling temperatures of these non-traditional constructional materials to at least 500°C.

CHAPTER 28

Lubricants and Detergents

MANY important phenomena are concerned with interactions at the boundaries of two different phases – at a liquid–liquid boundary or a liquid–solid boundary, for example. In this chapter we shall deal with two topics, of particular interest to engineers, where surface interactions are all-important, viz. lubrication and detergency.

Lubricants

28.1. The nature and mechanism of lubrication

Any two surfaces in contact with one another have to overcome frictional forces if they are to move with respect to one another. The work done in overcoming these forces generates heat and absorbs a considerable amount of power. It is the task of a lubricant to lessen frictional forces, and so lessen power losses and reduce heat output.

Lubricant action – the separation of two solid surfaces by a fluid in order to reduce shear stresses during motion – may differ from one situation to another. Three mechanisms are usually distinguished, and these we shall consider in turn.

Hydrodynamic lubrication, or fluid lubrication, occurs where there is a comparatively large separation of solid surfaces, i.e. of the order of 10^{-4} cm or greater. Given this condition, a continuous lubricant film can insert itself between the two surfaces. At the interface between solid and fluid the forces of attraction which exist cause an orientation of the fluid molecules, while within the main body of fluid the normal random distribution is maintained. This state of affairs is represented in Fig. 28.1. Polar molecules, substances like carboxylic acids and esters, attach themselves to solid surfaces more firmly than non-polar hydrocarbons and orientate more markedly, but their resistance to oxidation under hot working conditions is not favourable. A majority of hydrodynamic lubricants are petroleum-based hydrocarbons (containing, as we shall see, various additives).

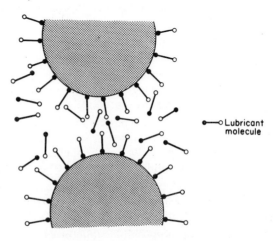

●—○ Lubricant molecule

FIG. 28.1. Hydrodynamic lubrication.

Boundary lubrication is necessary when two surfaces are in close contact with one another. In practice, two apparently smooth surfaces cannot be in intimate contact. On the molecular scale, even a highly polished surface has many irregularities and contact will only be made at odd points (points A in Fig. 28.2). If a long chain compound, polar at one end, non-polar at the other, is introduced between such surfaces, the polar end will again adhere strongly to the solid surface, while the rest of the molecule will orientate outward from the surface. Contact of the free non-polar end of one such molecule with a

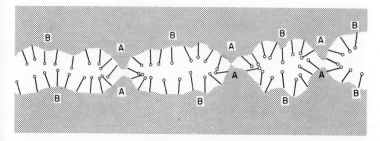

FIG. 28.2. Intermittent contact between two 'close' surfaces.

similar neighbouring molecule will produce low shearing stress conditions, i.e. will produce lubricating conditions. Thus carboxylic acids with long hydrocarbon chains, like stearic and oleic acid, have been used as boundary lubricants. Boundary lubrication (at points B in Fig. 28.2) will not prevent abrasion and wear occurring at contact points (A in Fig. 28.2).

Extreme pressure lubrication is a form of boundary lubrication designed to cope with high-temperature operating conditions. In this case solid films, typically phosphide films, are formed at the surfaces in question. These films, when subject to stress, melt and shear much more readily than the metal itself, and so lubrication is achieved. The films are regenerated *in situ* at the high operating temperatures by the continual presence in the lubricant medium of a suitable reagent. Tricresyl phosphate, $(H_3C \cdot C_6H_4 \cdot O)_3 \cdot OP$, is widely used for this purpose.

28.2. The viscosity of lubricants

A liquid becomes 'thinner' on heating. This change in viscosity with temperature is particularly important for lubricant oils, which often have to perform satisfactorily over considerable ranges of temperature.

Viscosity is a measure of the shear forces in a fluid. As these increase, i.e. as the viscosity of the oil increases, the work required to move a surface immersed in the oil will increase.

(Every motorist is aware of starting difficulties in winter weather, difficulties mainly due to the extra load produced by the cold viscous engine oil.) However, an oil which becomes too thin at higher temperatures may fail to withstand the pressure applied to it, and may be ejected from the regions between solid surfaces in close contact. Ideally one wants a lubricant whose viscosity does not change with temperature.

The viscosity of an oil may be measured and reported in a number of ways. Popular among these is to measure the flow time through a standard orifice; in the U.K. the Redwood viscosity is commonly quoted, in the U.S.A. the Saybolt viscosity is used. In these and other methods a given viscosity value will apply only at a particular temperature which should be quoted, i.e. a viscosity value in itself does not give any information about the variation of viscosity with temperature. This important relationship is indicated by the *viscosity index* (V.I.) of an oil, which is arrived at by comparing the change in viscosity of the test oil over a range of temperatures to that of two standard oils over the same range. The standard oils are given arbitrary V.I. values of zero (in the case of the oil with poor viscosity-temperature characteristics) and 100 (in the case of the oil with good viscosity-temperature characteristics). A high V.I. therefore denotes an oil which tends to exhibit minimal change in viscosity with changing temperature.

As with the octane rating of gasolines, the V.I. of a lubricating oil reflects the chemical structures present in the oil. For example, naphthenes (cycloalkanes) have lower V.I. values than alkanes. Again like the octane rating, the V.I. of an oil can be improved by various additives, and a V.I. in excess of 100 is now possible. Synthetic viscous polymers are used for this purpose, leading to the production of a range of visco-static lubricants. The widespread use of such oils has greatly lengthened engine life in motor cars.

The car owner is familiar with a designation of crankcase oil, which may be described by a coding such as SAE 20W or SAE 30 or some similar grouping. These designations represent another method of indicating viscosity values. They refer to divisions in the viscosity classification scheme set out by the Society of Automotive Engineers. The letter W indicates a winter grade oil.

28.3. Common lubricants

The structural requirements of a lubricant vary according to whether hydrodynamic, boundary, or extreme pressure lubrication is necessary. In the simplest case, that of pure hydrodynamic lubrication, almost any fluid will do – e.g. compressed air can be used. Seldom, however, is a clear-cut case met in practice, and a more likely system is one in which all three lubrication mechanisms are occurring at various points. Any lubricating medium in these circumstances must therefore meet a number of separate requirements.

Petroleum-based lubricants comprise the biggest class of lubricants. While a polar molecule adheres to a solid surface more firmly than a non-polar hydrocarbon, the latter has a greater resistance to oxidation and subsequent gum formation, and petroleum hydrocarbon fractions isolated from the higher boiling end of the petroleum distillate form the most important single source of lubricating materials. A 'simple' hydrocarbon lubricant in practice is likely to contain very small quantities of polar compounds which are thought to be important in the overall lubricating mechanism. (It is possible to refine too highly and lessen the effectiveness of a lubricating oil.) Widely used are *compounded oils*, which consist of a hydrocarbon fraction blended with polar oils derived from animal or vegetable sources. Lubricant *greases* are similar, being a blend of a hydrocarbon and a soap. We shall discuss these polar oils and soaps when we look at the nature of detergent materials. Other varieties of lubricants are the *cutting oils* used in production engineering; these oils, often milky white in appearance, are emulsions of hydrocarbon and polar oil in aqueous soap solution, the oil medium acting as a good lubricant, the aqueous medium with its high specific heat and thermal conductivity absorbing much of the heat generated in the process.

Synthetic lubricants are varied in type. Hydrocarbon polymers like polyethylene and polybutylene are used either as direct substitutes for petroleum lubricants, or as additives intended to raise the viscosity index of the oil. Synthetic polymers containing polar groups are used where special features such as low volatility or low temperature operation are required. An interesting range of inorganic lubricants is provided by *silicone oils*, which are expensive, but have outstanding properties in terms of high viscosity indices and high thermal stabilities. *Fluorinated* and *chlorinated hydrocarbons* are useful where a risk of inflammability exists or where a high degree of chemical stability is required.

Solid lubricants have long been in use. (The distinction between solid and fluid lubricants may be more apparent than real. It has been demonstrated that fluids under very high pressure, as at an interface, exhibit very great increases in effective viscosity and behave as solids.) *Graphite* is the best known example. Its lubricant properties were once solely attributed to its layered structure (Section 6.2) in which one graphite plane could slip easily with respect to another, but it has been demonstrated that gases adsorbed on the graphite surface are also necessary. Graphite is used as a component in oil and grease mixtures, sometimes being dispersed in the oil as a colloid. In this latter state it tends to cling to metal surfaces more readily than a purely fluid lubricant, and is resistant to oxidation. *Molybdenum disulphide*, which also has a layer type of structure, is used as a high-temperature lubricant, and as an additive to oils. The disulphide adheres to the solid surface, producing, it is thought, a low coefficient of friction because of the subsequent ease of shearing of molybdenum–sulphur bonds.

28.4.   Additives

Lubricants contain a number of additives apart from those which themselves contribute to the lubrication mechanism. Engine oils normally have present a *detergent*, the purpose of which is to free the surfaces from dirt and carbonaceous deposits. *Antioxidant* additives are essential components of many organic preparations, including lubricating oils; amine compounds are commonly used to inhibit oxidation reactions. To minimize the effect of crystallization of heavier hydrocarbons from the fluid at low working temperatures a *pour point depressant* is added to lubricant oils. This does not necessarily prevent crystallization occurring, but is ensures that the crystals present are much smaller and less troublesome than would otherwise be the case. Lastly, *corrosion inhibitors* are added to protect metal bearings.

Soaps and Detergents

28.5. Detergent action

The process of deterging is one of dirt and grease removal from a surface. The agents used to bring this about are called soaps and synthetic detergents ('syndets' in industrial jargon). Dirt is normally in itself partly or solely organic and therefore not readily soluble in or removed by water, and a detergent must be employed.

Detergent action is complex, but we can simplify it by dividing it into two distinct steps. The first of these is to bring the detergent solution into intimate contact with the surface to be cleaned, a process called *wetting*; the second is to remove the dirt from the surface region by *emulsification* in the aqueous phase.

Both wetting and emulsification mechanisms are dependent on the chemical structure of the detergent molecules. A typical structure consists of a long hydrocarbon chain terminating in a highly polar function. Thus the sodium salt of a long-chain carboxylic acid like stearic acid, or the quaternary ammonium salt of a long chain amine like cetyltrimethylamine both have detergent properties.

$$C_{17}H_{35}{\cdot}CO_2{}^-Na^+ \qquad C_{16}H_{33}{\cdot}\overset{+}{N}(CH_3)_3Cl^-$$

Sodium stearate   Cetyltrimethylammonium
chloride

The hydrocarbon part of these compounds, which is typically organic, covalent, and therefore non-polar, has no affinity for or tendency to dissolve in water and is termed the *hydrophobic* group. The polar part, conversely, will freely dissolve in water to give ionic species, and is termed the *hydrophilic* group. Every detergent contains in some form two such groups. In a two-phase system, a molecule will tend to orientate itself at the interface such that the hydrophobic group is in the organic phase, the hydrophilic end in the aqueous phase. This ordered arrangement at the interface lowers the surface tension in that region. It is this lowering of interfacial energy which constitutes the driving force in the wetting and emulsification processes.

Let us consider a surface covered with an oil film. Water will be repelled by the oil film in this system, and wetting will not occur. If a detergent is present, however, the hydrophobic group can penetrate the oil film, and become attached to the surface proper. Adsorbtion of the detergent at the surface displaces the oil, which rolls up to form a globule. This 'rolling-off' phenomenon is beautifully illustrated in Fig. 28.3.

The oil globules, having formed, are removed by mechanical agitation and subsequent emulsification in the aqueous medium. In this step, the hydrophobic groups of the detergent molecules orientate towards the oil globules, while the ionized hydrophilic ends are directed outwards into the surrounding aqueous medium. Each oil globule therefore acquires a shield of charged particles. One globule will be repelled by another, coagulation will be prevented, and dispersion in the aqueous phase encouraged. This is represented in Fig. 28.4 in which the action of, for example, sodium stearate is seen to produce a negatively charged anionic species at the emulsified particle.

The picture outlined above, while simplified, contains the essential ingredients of detergent action. Instinctively we tend to associate a lather with good cleansing properties. A decrease

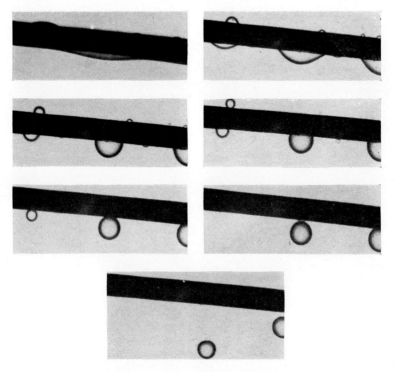

FIG. 28.3. Detergent action at a fibre surface. (Reproduced by permission of Unilever Research.)

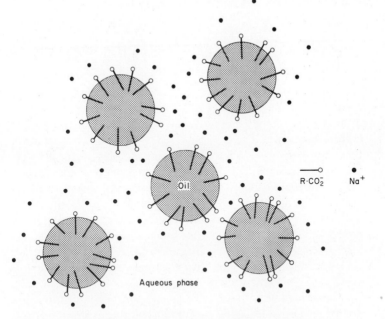

$$\overline{\hspace{2cm}}\!\!-\!\!o \quad R \cdot CO_2^- \qquad \bullet \quad Na^+$$

Fig. 28.4. Mutual repulsion of emulsified globules.

in surface tension (such as occurs in a detergent solution) will always enhance the formation of bubbles, but these are not thought to be a necessary feature of a cleansing solution. Dirt may be entrained by and carried away in the foam, but this is a secondary effect.

28.6. Soaps

Soaps are detergents derived from naturally occurring oils, fats and waxes which in turn are available from vegetable sources. These oils are glycerides, i.e. they are esters of the trihydric alcohol glycerol and a number of long-chain carboxylic acids. The hydrolysis of a glyceride, a reaction called *saponification*, yields glycerol and a mixture of salts of the respective acids:

$$
\begin{array}{l}
CH_2 \cdot O \cdot OC \cdot R \\
| \\
CH \cdot O \cdot OC \cdot R' \quad + 3NaOH \longrightarrow \\
| \\
CH_2 \cdot O \cdot OC \cdot R''
\end{array}
\qquad
\begin{array}{ll}
CH_2 \cdot OH & R \cdot CO \cdot ONa \\
| \\
CH \cdot OH & + R' \cdot CO \cdot ONa \\
| \\
CH_2 \cdot OH & R'' \cdot CO \cdot ONa
\end{array}
$$

Glyceride   Glycerol   Sodium salts
                                               (Soap)

Some of the more common acid structures occurring in glycerides have been listed in Table 23.2, from which we can see that they consist of a long hydrocarbon chain (typically containing fifteen or seventeen carbon atoms, and perhaps containing one or more carbon–carbon double bonds) terminating in a polar carboxylic acid group. The sodium or potassium salts, with a polar hydrophilic group attached to a hydrophobic hydrocarbon group, have the necessary structural features required of a detergent, and are called *soaps*.

Glycerides, which vary from simple esters, in which the same acid fragment occurs at all three ester linkages, to complex esters, which yield mixed acids on saponification, are important not only in the soap industry but also in foodstuffs (animal and vegetable fats are glycerides essentially) and in the manufacture of paints, varnishes and linoleum. The alcohol component of saponification, glycerol, has several outlets which include its use as an anti-freeze, as a source of synthetic polyesters, and as the stock material for nitroglycerin manufacture; demand for glycerol is such that much has also to be synthesized from propylene.

28.7. Synthetic detergents

Soaps have a number of disadvantages. In hard water the carboxylic acid anion precipitates out as insoluble calcium and magnesium salts and becomes ineffective; soaps tend to be sensitive to acid conditions; moreover, the demand for soaps has outstripped the supply of raw materials. Synthetic detergents, which overcome these disadvantages, have become very important and now account for well over half of the detergent market.

The largest single class of synthetic detergents is that in which the hydrophilic group is the sodium salt of a sulphonic acid or sulphonated alcohol. The sulphonic acid type is prepared by alkylating the benzene ring with a group containing ten or twelve carbon atoms, and sulphonating the nucleus:

| Alkyl benzene | Alkyl benzene sulphonate | Alkyl benzene sulphonate (Sodium salt) |

Alcohols for sulphonation are prepared by catalytic reduction of long-chain acids derived from glycerides, or by synthesis from alkenes. If in these structures the group R is branched,

$$R \cdot CO_2H \longrightarrow R \cdot CH_2 OH \xrightarrow[Na_2CO_3]{N_2SO_4} R \cdot CH_2 \cdot O \cdot SO_3Na$$
Alkenes

Long-chain alcohol

Alcohol sulphate (Sodium salt)

resistance to degradation by biochemical organisms is high. Such biologically 'hard' detergents have in the past caused considerable problems of river pollution, and are now avoided.

Sulphates and sulphonates ionize in water to yield species in which the anion, $RO \cdot SO_3^-$ or $R \cdot SO_3^-$, is the active participant in the detergent process. *Anionic detergents* are by far the biggest class.

Cationic surface active agents are more often used as textile softeners and as germicides than as detergents. Quaternary ammonium salts (chlorides or bromides), whose structure may be written as $R_4N^+Cl^-$, typify this class.

Some surface active agents are *amphoteric*, containing both anionic and cationic species. They take the form of the sodium salt of an amino acid, i.e. $R \cdot NH \cdot (CH_2)_2 \cdot CO_2Na$.

Nonionic detergents are also produced, and indeed are much more important than cationic and amphoteric types. Here interfacial activity depends on the presence of a number of groups, polar in nature and therefore having an affinity for water, but not ionizable in water. Long-chain ethers, made by reaction of alcohols or phenols with ethylene oxide, provide an important

$$R - \underset{\text{Alkyl phenol}}{\bigcirc} - OH \quad + \quad \underset{\substack{\text{Ethylene} \\ \text{oxide}}}{n \cdot CH_2 - CH_2} \longrightarrow R - \underset{\text{Long-chain polyether}}{\bigcirc} - (O \cdot CH_2 \cdot CH_2)_n \cdot OH$$

example. The interaction with water is thought to occur through hydrogen bonding at the ether oxygen.

$$\underset{\text{Ether link}}{-O-} \quad + \quad H_2O \longrightarrow \underset{\substack{H - O - H \\ \text{Hydrogen bonded ether}}}{-O-}$$

28.8. Detergent builders

The efficiency of detergents is improved by the presence of additives known as detergent builders. These are electrolytes, the most common of which at the present time are the sodium phosphates. One of the main functions of these substances, whose role in the detergent process is recognized as complex, is to encourage the emulsification of greases.

Table of Elements, Symbols, Atomic Numbers, Atomic Weights

Name	Symbol	Atomic number	Atomic weight	Name	Symbol	Atomic number	Atomic weight
Actinium	Ac	89	(227)	Mercury	Hg	80	200·59
Aluminium	Al	13	26·98	Molybdenum	Mo	42	95·94
Americium	Am	95	(243)	Neodymium	Nd	60	144·24
Antimony	Sb	51	121·75	Neon	Ne	10	20·183
Argon	Ar	18	39·948	Neptunium	Np	93	(237)
Arsenic	As	33	74·92	Nickel	Ni	28	58·71
Astatine	At	85	(210)	Niobium	Nb	41	92·91
Barium	Ba	56	137·34	Nitrogen	N	7	14·007
Berkelium	Bk	97	(249)	Nobelium	No	102	(253)
Beryllium	Be	4	9·012	Osmium	Os	76	190·2
Bismuth	Bi	83	208·98	Oxygen	O	8	15·999
Boron	B	5	10·81	Palladium	Pd	46	106·4
Bromine	Br	35	79·909	Phosphorus	P	15	30·974
Cadmium	Cd	48	112·40	Platinum	Pt	78	195·09
Caesium	Cs	55	132·91	Plutonium	Pu	94	(242)
Calcium	Ca	20	40·08	Polonium	Po	84	(210)
Californium	Cf	98	(251)	Potassium	K	19	39·102
Carbon	C	6	12·011	Praseodymium	Pr	59	140·91
Cerium	Ce	58	140·12	Promethium	Pm	61	(147)
Chlorine	Cl	17	35·453	Protactinium	Pa	91	(231)
Chromium	Cr	24	52·00	Radium	Ra	88	(226)
Cobalt	Co	27	58·93	Radon	Rn	86	(222)
Copper	Cu	29	63·54	Rhenium	Re	75	186·23
Curium	Cm	96	(247)	Rhodium	Rh	45	102·91
Dysprosium	Dy	66	162·50	Rubidium	Rb	37	85·47
Einsteinium	Es	99	(254)	Ruthenium	Ru	44	101·1
Erbium	Er	68	167·26	Samarium	Sm	62	150·35
Europium	Eu	63	151·96	Scandium	Sc	21	44·96
Fermium	Fm	100	(253)	Selenium	Se	34	78·96
Fluorine	F	9	19·00	Silicon	Si	14	28·09
Francium	Fr	87	(223)	Silver	Ag	47	107·870
Gadolinium	Gd	64	157·25	Sodium	Na	11	22·9898
Gallium	Ga	31	69·72	Strontium	Sr	38	87·62
Germanium	Ge	32	72·59	Sulphur	S	16	32·064
Gold	Au	79	196·97	Tantalum	Ta	73	180·95
Hafnium	Hf	72	178·49	Technetium	Tc	43	(99)
Helium	He	2	4·003	Tellurium	Te	52	127·60
Holmium	Ho	67	164·93	Terbium	Tb	65	158·92
Hydrogen	H	1	1·0080	Thallium	Tl	81	204·37
Indium	In	49	114·82	Thorium	Th	90	232·04
Iodine	I	53	126·90	Thulium	Tm	69	168·93
Iridium	Ir	77	192·2	Tin	Sn	50	118·69
Iron	Fe	26	55·85	Titanium	Ti	22	47·90
Krypton	Kr	36	83·80	Tungsten	W	74	183·85
Lanthanum	La	57	138·91	Uranium	U	92	238·03
Lawrencium	Lw	103	(257)	Vanadium	V	23	50·94
Lead	Pb	82	207·19	Xenon	Xe	54	131·30
Lithium	Li	3	6·939	Ytterbium	Yb	70	173·04
Lutetium	Lu	71	174·97	Yttrium	Y	39	88·91
Magnesium	Mg	12	24·312	Zinc	Zn	30	65·37
Manganese	Mn	25	54·94	Zirconium	Zr	40	91·22
Mendelevium	Md	101	(256)				

PERIODIC TABLE OF THE ELEMENTS

Period	Valence Shell	IA s^1	IIA s^2	IIIB $d^1s^2f^x$	IVB d^2s^2	VB $(d^3s^2)‡$	VIB $(d^5s^1)‡$	VIIB d^5s^2	VIII $(d^6s^2)‡$	VIII $(d^7s^2)‡$	VIII $(d^8s^2)‡$	IB s^1d^{10}	IIB s^2	IIIA s^2p^1	IVA s^2p^2	VA s^2p^3	VIA s^2p^4	VIIA s^2p^5	0 s^2p^6
n=1	1s	1 H 1.00797																	2 He 4.0026
n=2	2s2p	3 Li 6.939	4 Be 9.0122											5 B 10.811	6 C 12.01115	7 N 14.0067	8 O 15.9994	9 F 18.9984	10 Ne 20.183
n=3	3s3p	11 Na 22.9898	12 Mg 24.312											13 Al 26.9815	14 Si 28.086	15 P 30.9738	16 S 32.064	17 Cl 35.453	18 Ar 39.948
n=4	4s3d4p	19 K 39.102	20 Ca 40.08	21 Sc 44.956	22 Ti 47.90	23 V 50.942	24 Cr 51.996	25 Mn 54.9380	26 Fe 55.847	27 Co 58.9332	28 Ni 58.71	29 Cu 63.54	30 Zn 65.37	31 Ga 69.72	32 Ge 72.59	33 As 74.9216	34 Se 78.96	35 Br 79.909	36 Kr 83.80
n=5	5s4d5p	37 Rb 85.47	38 Sr 87.62	39 Y 88.905	40 Zr 91.22	41 Nb 92.906	42 Mo 95.94	43 Tc 99	44 Ru 101.07	45 Rh 102.905	46 Pd 106.4	47 Ag 107.870	48 Cd 112.40	49 In 114.82	50 Sn 118.69	51 Sb 121.75	52 Te 127.60	53 I 126.9044	54 Xe 131.30
n=6	6s4f5d6p	55 Cs 132.905	56 Ba 137.34	57–71 ★	72 Hf 178.49	73 Ta 180.948	74 W 183.85	75 Re 186.2	76 Os 190.2	77 Ir 192.2	78 Pt 195.09	79 Au 196.967	80 Hg 200.59	81 Tl 204.37	82 Pb 207.19	83 Bi 208.980	84 Po 210	85 At 210	86 Rn 222
n=7	7s5f6d7p	87 Fr 223	88 Ra 226	89–103 ★															

Representative Elements · Transition Elements—d · Representative Elements · Noble Gas Elements

Inner Transition Elements—f

★ Lanthanide Series	57 La 138.91	58 Ce 140.12	59 Pr 140.907	60 Nd 144.24	61 Pm 145	62 Sm 150.35	63 Eu 151.96	64 Gd 157.25	65 Tb 158.924	66 Dy 162.50	67 Ho 164.930	68 Er 167.26	69 Tm 168.934	70 Yb 173.04	71 Lu 174.97
★ Actinide Series	89 Ac 227	90 Th 232.038	91 Pa 231	92 U 238.03	93 Np 237	94 Pu 242	95 Am 243	96 Cm 247	97 Bk 247	98 Cf 249	99 Es 254	100 Fm 253	101 Md 256	102 No 254	103 Lw 257

‡Variable valence shells.

412

Index